Protected Areas

Josephine Gillespie

Protected Areas

A Legal Geography Approach

palgrave
macmillan

Josephine Gillespie
School of Geosciences
University of Sydney
Sydney, NSW, Australia

ISBN 978-3-030-40501-4 ISBN 978-3-030-40502-1 (eBook)
https://doi.org/10.1007/978-3-030-40502-1

This Palgrave Pivot imprint is published by the registered company Springer Nature Switzerland AG
The registered company address is: Gewerbestrasse 11, 6330 Cham, Switzerland

Preface

The purpose of this book is to rethink protected area management to take into account the importance of getting the law right. It argues that a thorough consideration of the people-place-law nexus must become a priority consideration in conservation policy making. On the brink of an extinction crisis, today's rates of biodiversity loss demand effective in situ conservation through protected area programs. Improving the effectiveness of protected areas requires a re-think about current approaches. I argue that taking into account existing people-place-law dynamics must not be a secondary consideration after the protected area is created. Too often these considerations are not prioritised in protected area designations.

My previous experience in the legal profession informs my approach. After many years in litigation I came to the belief that laws must be useful; that they must achieve what they are written for. I also believe that laws must be written for the people and places that they will impact. To write and implement laws in isolation from their context leads to unsatisfactory outcomes. Since returning to the university sector I now teach and research in the interdisciplinary area of law and geography. I turn to the scholarship of legal geography to implement this perspective.

Legal geography lies at the heart of this book. The key to improving current protected area approaches lies in integrating natural and social sciences. Understanding people-place dynamics is essential to this task. In doing this we need to consider the role of law in both shaping and reacting to people-place dynamics. Too often the economic, political and social are considered, but the law can, and often does, function as an independent structural component in governance decision-making. I argue that the law should not be overlooked as an independent factor in shaping people-place dynamics. This book seeks to recalibrate the role of law as an essential element in people-place dynamics. Understanding this role will enhance current protected area management practices.

I acknowledge that parts of this book are simplified to provide an overview of complex issues. The book is designed to synthesise rather than extrapolate complex conservation practice conundrums. I have not sought to achieve extensive coverage of the issues presented in this book but I have sought to introduce the idea that focusing on people-place-law dynamics is critical in driving new, fresh ways of thinking about how to make our important in situ conservation protected areas better.

Sydney, Australia Josephine Gillespie

Acknowledgements

I would like to thank my colleagues in the field of legal geography. Within the Institute of Australian Geographers Legal Geography Study Group are a core group of cognate researchers who collegiately support the evolution of the dynamic field of legal geography. Without this stimulating group, our Australian-influenced brand of legal geography scholarship would be considerably poorer. Light-hearted moments are important to me and I would like to recognise my colleagues in the School of Geosciences at the University of Sydney for their support and encouragement, and most of all, lots of laughter. I also want to acknowledge the importance of successive undergraduate student cohorts and especially my recent Honours students who always challenge and inspire my research. Finally, I want to thank my family and particularly Angus, Will and Dan, who always provide me with practical, sensible and down-to-earth advice.

Contents

Abbreviations/Acronyms

ICOMOS	International Council on Monuments and Sites
IPBES	Intergovernmental Science-Policy Platform on Biodiversity and Ecosystem Services
IUCN	International Union for Conservation of Nature
OECM	Other Effective Area-Based Conservation Measures
TWWHA	Tasmanian Wilderness World Heritage Area
UNEP	United Nations Environment Programme
UNESCO	United Nations Educational, Scientific and Cultural Organization
WCMC	World Conservation Monitoring Centre
WCPA	World Commission on Protected Areas
WPDA	World Database on Protected Areas
WWF	World Wildlife Fund
ZEMP	Zoning and Environmental Management Plan

List of Tables

CHAPTER 1

Challenges for Protected Areas: Biodiversity Loss, Place-People and Law Connections

Abstract In this chapter, I consider the extraordinary phenomenon of planetary-scale biological diversity loss and introduce how one key answer aimed at halting these losses, namely in situ protected area conservation programs, are failing. Despite unprecedented efforts to preserve ecosystems and habitats we are facing an extinction crisis. In this context I introduce a new approach aimed at improving the efficacy of in situ conservation protected area schemes. I argue one solution can be found in legal geography scholarship. A legal geography approach empowers us to consider the intertwined regulatory people-place dynamics that enable or disable effective management practices. In the age of the Anthropocene, legal geography is both an applied and conceptual approach for our time.

Keywords Biodiversity loss · Protected areas · In situ conservation · Legal geography

1.1 Introduction: The Protection Imperative

The extent and degree of species and habitat losses across the globe are staggering. Few would not be shocked knowing that the current extent of species loss is comparable to global extinction events of the geological past. We may now be in the midst of a sixth global extinction event (Barnosky et al. 2011). In 2019 the Intergovernmental Science-Policy

J. Gillespie, *Protected Areas*,
https://doi.org/10.1007/978-3-030-40502-1_1

Platform on Biodiversity and Ecosystem Services (IPBES 2019) estimates that there are more than one million species at risk of extinction (following Mora et al. 2011). The 2018 Living Planet Report, produced by the World Wildlife Fund and the Zoological Society of London, reports a 60% decline of animal populations across the globe between 1970 and 2014 driven by habitat loss and degradation (WWF 2018).

Significant academic work also demonstrates the horrific trajectory of loss (see for example Steffen et al. 2015; Johnson et al. 2017). Despite the dedicated initiatives to stem biodiversity loss and the degradation of habitat, decline continues. There are widespread and aggressive ecosystems losses globally (for example, see Watson et al. 2018), with rates of terrestrial wilderness lost to, inter alia, human settlement and farming between 1993 and 2009 are estimated to be 3.3 million square kilometres (Watson et al. 2016a, b). Inevitably, we must question whether global conservation policy is doing enough to slow biodiversity loss and environmental degradation (Pimm et al. 2018; Mappin et al. 2019). We must also question whether the protected area approach is succeeding as well as it might (Jones et al. 2018; Watson et al. 2016a, b, 2018). In this context, the rationale for protected areas has shifted from the romanticised ideals of nature that underwrote the first national parks to an acute focus on the integrity of ecosystem function and toward global-scale biogeochemical cycles. Assessing the success of protected areas is critical in this context (Joppa et al. 2016). Di Marco and colleagues (2019), using a modelling approach, argue that the retention of dedicated areas for the protection of habitat is essential;

> … retaining these remaining wilderness areas is essential for the international conservation agenda. Wilderness areas act as a buffer against species loss, as the extinction risk for species within wilderness communities is—on average—less than half that of species in non-wilderness communities. (p 582)

The authors call for targeted protection to ensure areas of high habitat value for biodiversity receive more (rather than less) protection.

The protection imperative lies at the heart of a trillion-dollar global industry. The estimated value of global ecosystem services in 2011 was US$125 trillion (Costanza et al. 2014). Setting aside the merit of purely economic valuation of ecosystem services, or 'nature' writ large, there is no doubt that biological diversity loss requires comprehensive and effective policy responses.

Attempts to improve biodiversity conservation outcomes through the refinement of protected area approaches is widespread amongst academics and practitioners alike. The World Database on Protected Areas (WDPA) is updated monthly and is a publicly available online dataset maintained by leading global biodiversity experts, especially the United Nations Environment World Conservation Monitoring Centre with support from the International Union for the Conservation of Nature (IUCN) and its Commission on Protected Areas (UNEP and WCMC undated). The work of the IUCN on protected areas (for example, see Lausche and Burhenne Guilmin 2011) steams ahead through their 'Global Standard' program which aims to set a standard for assessment of protected area outcomes (including for equitable governance and management [IUCN and WCPA 2017]).

1.2 A Question of Control

Adams and Hutton (2007) reminds us that conservation is about managing people, not plants or animals. Habitat loss as a result of the activities of people within and around protected areas, remains the primary driver of species extinction, even in the context of planetary-scale climatic change (Sala et al. 2000). The success of protected areas is, therefore, as much a matter for social scientists as it is for natural scientists. Separating an extraordinary nature from the lived-in ordinary world is short-sighted (Olwig 1995). Eliminating the binary is an immediate demand for sustainable futures. Just as Whittaker et al. (2005) called for greater engagement within the natural sciences for evaluation of protected area planning frameworks under the rubric of 'conservation biogeography', there is now a greater awareness of the socio-legal setting within which protected areas operate. Indeed, since Agrawal and Ostrom (2006) famously called the interaction between social scientists and conservation biologists "dialog of the deaf", there has been an expansion of focussed inquiry into this critical interface (Adams and Hutton 2007; Treves et al. 2006; Chan et al. 2007; Bennett et al. 2017; Blicharska et al. 2016 and many others, see also Mascia et al. 2003).

Law, in all of its various guises, is the vehicle by which humans are regulated yet, oftentimes, law is overshadowed or abstracted in this literature by a focus on social/cultural, political or economic factors. Law can be, perhaps especially in a Western context, independent from political and economic considerations through the separation of powers doctrine. Yet I argue that an independent judiciary can play a significant part in shaping people/place dynamics (see also O'Donnell 2020). This separation makes the focus on regulatory dimensions important to think through as a distinctive component within environmental policy and decision-making. Moreover, entangled within the 'human factor' are questions about legitimacy, accountability and transparency in decision-making that sit squarely within legal processes (Brechin et al. 2002). There is a need to fully understand the importance of the regulatory framework—from drafting to implementation—to improve conservation outcomes in protected areas.

This necessary understanding of the regulatory frameworks cannot be achieved, I argue, without an attention to the idiosyncrasies of place, which has traditionally been the focus of geographers and cognate social science researchers. Together, critical attention to both law and place—and particularly the interdependencies between them—enables better-fitting solutions for communities living in or adjacent to protected areas, thus increasing the effectiveness of protected areas in mitigating biodiversity loss and environmental degradation.

Adams' recent observations (2019) drawing our attention to the essential work of geography and the spatiality of conservation still does not engage as it might with the idea of law as an independent force or driver of the conservation agendum. For Adams (2019) law becomes a tool of restriction (demarcating uses) and enforcement (dealing with breaches)—which are legitimate elements in need of greater attention in conservation management practice—but which are not a full picture in considering law's remit in conservation. An explicit focus on the intertwining of place and law should consider law in all its guises.

The need for this research is particularly acute in the Global South, where 'top down' protected area schemes emerging from the Global North, and Anglo-American legal tradition, are often grafted onto existing (frequently normative) regulatory systems that may be quite different (Gillespie 2018). In such circumstances, adherence to and enforcement of protected area regulations may be extremely problematic.

1.3 A New Approach: Legal Geography Perspectives

The value of law and geography scholarship, I argue, lies in the refreshing perspective that comes from acknowledging the reciprocal, dependent relationship between geographical concepts of place and space, and legal concepts of regulation, fairness and justice. Through laws, spaces become controlled and regulated. The physical form of place may change as a consequence, and so law becomes codified into the material of landscape as an enduring legacy that exerts itself on the present and future. Yet little consideration has been paid to the way that legal apparatus work *through* local laws or regulations to create these controls. Legal geography, then, seeks to determine, inter alia, whether regulations are well understood by those affected by them, whether regulations have a positive or negative impact on existing land use practices/livelihoods, and the extent to which collaboration or consultation has been achieved in framing conservation/preservation approaches.

Specifically, I argue a legal-geographical lens can be comprised of the following four key elements:

1. Translation of law through layers (as in global, national, regional, local and hierarchical or spatial/territorial) from centralised to decentralised places or vice versa;
2. Unpacking the relationship between formal law and informal law, which often time takes the form of lore. This requires consideration of pre-existing customs, traditions and normative everyday practices of people and places;
3. Accounting for visible and invisible relational governance and institutional functions (including case law and the independent role of the judiciary, especially in a Western setting) and interwoven complexities/interactions for a range of issues which brings into focus regulatory/people dynamics, especially as opposed to focussing in on economic, political or cultural elements;
4. A focus on the ambiguous role of place. Following Massey (2005) whose work draws our attention to the problematics associated with place analysis: "(p)lace plays an ambiguous role ... Horror at local exclusivities sits uneasily against support for the vulnerable struggling to defend their patch" (p. 6).

Together these elements both underpin the value of legal geography as a method (see O'Donnell et al. 2020) and as an analytical approach. In any contemporary analysis that seeks to critique dominant paradigms, especially in efforts to re-consider Western power dynamics and thus to refocus attention towards de-colonisation and more-than-human demands, legal geography becomes a powerful analytical lens.

The legal geography lens also allows us to unpack those things that are not readily observable, particularly to 'outsiders' (element 3, to make visible the invisible, following Bartel et al. 2013). Unpacking the dynamics of place-based power and decision-making processes can reveal intangible hierarchies of administrative power that are difficult to document from a removed empirical perspective, but which are more suitable to an engaged qualitative approach. Deeply embedded relational interactions revealing agency and contingent practices can be unpacked through in-depth qualitative research practices for which legal geography scholars are increasingly known (see O'Donnell et al. 2020; Orzeck and Hae 2019).

1.4 Book Structure

In this opening chapter we consider the big picture question: why we should we be concerned with thinking about conservation of species and habitats through a new lens. The crippling reality of biodiversity loss and our attempts to redress these losses pushes us into action—we now protect more parts of the planet than ever before. Yet, the system is not working as well as it might. In these circumstances I suggest we turn to legal geography as an approach to redefine how we manage protected areas to secure habitats and ward off species losses. Chapter 2 introduces in more detail the protected area phenomenon. Chapter 2 is structured in three parts. The first section briefly delves into the history and evolution of the protected area idea and form. This section enables us to understand how the idea of a dedicated bounded and zoned location evolved in a New World context. The protected area ideal was a product of a time and place, yet it continues to hold fast and is very much cemented in global conservation agenda—it is an essential ingredient in conservation efforts throughout the world. Protected areas' central position is intimately linked to today's biological diversity extinction crises. In the second section of this chapter we consider the biodiversity conservation imperative. In the age of the Anthropocene the dramatic loss of biodiversity across the globe is profound. Biological diversity losses are severe, with estimates that more than 1 million species are at risk of extinction. Conservation efforts to stem the gargantuan losses are everywhere. In place (in situ) conservation to retain habitat and ecosystem functioning is crucial in efforts to prevent catastrophic loss of biodiversity. Despite incredible efforts through habitat protection to stem losses we need more. Here we turn to think about whether different approaches might improve the current protected area governance efforts. In this part of Chapter 2, I turn to consider the value of adopting a legal geography lens to forge new direction in biodiversity conservation research efforts.

From Chapter 2 we know that the protected area approach faces many challenges and has not been as successful as it might have been. In Chapter 3 we consider applying a different lens to our analysis of environmental protection through protected areas. A legal geography approach can sharpen our focus upon the shortcomings in the current approach and provide a path for positive change. This chapter makes the case that legal geography can make a significant contribution to the environmental regulation of protected area habitats. We consider the importance of place and material considerations in our analysis of protected area regulatory regimes and I turn our attention towards the importance of pre-defined concepts linked to property and territory in how we approach access and use of land and water scapes. In this chapter, I also urge us to consider the conceptual linkages between scholarship in the area of natural resource management and environmental protection. Here the associated scholarship of political ecology and environmental/ecological justice come to the fore and I argue for a different inflection: a legal ecological approach. Approaching the problem of biodiversity loss, habitat protection, competing stakeholder demands and changing human/more-than-human interactions in the age of the Anthropocene demands that we use a legal geography lens that sees beyond the dominant paradigm. The final part of this chapter encourages us to think outside the dominant paradigm to forge new pathways that embrace change.

Chapter 4 delves into an examination of the World Heritage protected area regime. This chapter, together with Chapter 5 dedicated to exploring Ramsar protected wetlands, offers insights into the particularities of two different international protected area governance regimes. A focus on these particular protected area regimes holds currency as the IUCN, a leading global environmental organisation, has been investigating the rationale for harmonising different types of protected areas regimes, including both Ramsar wetlands and World Heritage properties (Schaaf and Rodrigues 2016). Chapter 4 provides an outline of the history leading up to the creation of the *World Heritage Convention* to understand the way in which this particular protected area regime came into being, and to highlight the foundations upon which this very successful international environmental protection regime is built. The next section in this chapter then considers the Australasian regional context of world heritage property management, using selected examples from Southeast Asia (Angkor, Cambodia and Trang An Landscape Complex, Vietnam), the Pacific/Oceanic region (Chief Roi Mata's Domain, Vanuatu and East Rennell, Solomon Islands) and Australia (the Tasmanian Wilderness World Heritage Area and the Great Barrier Reef, Queensland). Although these settings are vastly different in their physical locations and traits, and in their cultural contexts, the challenges besetting management are eerily similar. Our focus on the connections between people/place and regulatory dynamics draw attention to the shortcomings in our regulatory approaches—for places of outstanding universal value.

Chapter 5 considers another particular protected area type: *Ramsar* wetlands. Wetland habitats are amongst the most ecologically and biologically diverse on the planet. Wetlands hold intrinsic value as sites of diversity, but they are also areas of enormous anthropocentric value—as carbon storage in an era of increasing greenhouse gas production and accelerated global warming. Human connections to wetlands are complex and, in this chapter, we first turn to look at the perceptions within popular culture of wetland areas across the globe—as sites of despair and melancholy to sites of regeneration. Despite the complexity of the people/place wetland connections, and the enormous value of these sites, global wetland destruction is unprecedented in the past 100 years. Whether by reclamation for urban expansion, conversion for agricultural purposes or alteration for recreational uses, the rate of wetland loss is severe. In this chapter we articulate the extent and rate of losses. I then turn to consider the management of selected wetlands in the Australasian region. We begin this journey in the great lake, or Tonle Sap, in central Cambodia, in mainland Southeast Asia. The Tonle Sap is the largest freshwater lake in Southeast Asia and supports the livelihoods and diets of millions of people. The Tonle Sap wetlands, however, are under threat. Changing agricultural practices within the wetlands and flooded forests of the lake's edge mark the demise of important waterbird sanctuaries. The implications for biodiversity are unclear. What is clear, however, are the challenges facing communities living and working in and around these wetlands. Our next case study comes from wetlands scattered throughout the Oceania region—an underrepresented area for Ramsar wetland sites. Wetlands in these locations hold no less value than other such sites, yet the few sites stand as testimony to the lack of value placed on this ecosystem type. Challenges through progressive regulatory regimes which pay attention to customary practices also beset this protected area. In the final regional setting, we turn to consider Ramsar wetlands in Australia using the coastal lakes of the lower Murray River system as our case study. Here we see that despite the existence of laws designed to protect wetlands a variety of pressures can adversely impact environmental objectives. Standing back we can see that in all regional wetlands examples there exists a dissonance between place, people and the wetland protection laws.

Finally, in Chapter 6, I provide some concluding commentary about the state of protected areas. Legal standards sit at the forefront of improving conditions for all. Without regulatory intervention places would be vastly different. Environmental protection regimes work. Restricting human uses enables more-than-human communities to exist—giving them agency and access to resources to sustain their populations. Law is not immune from criticism, however, and laws that are not sufficiently linked to place-conditions can prove costly. This, of course, is the value of a legal geography approach. Thinking through the connections between physical landscapes, the people who live and work within these places, and the ways to regulate to achieve multiple objectives and to seek compromise with promise lies at the heart of this endeavour.

References

Adams WM and Hutton J (2007) People, parks and poverty: Political ecology and biodiversity conservation. *Conservation and Society* 5(2): 147.

Adams WM (2019) Geographies of conservation III: Nature's spaces. *Progress in Human Geography*. Available https://doi.org/10.1177/0309132519837779.

Agrawal A and Ostrom E (2006) Political science and conservation biology: A dialog of the deaf. *Conservation Biology* 20(3): 681–682.

Barnosky AD, Matzke N, Tomiya S, Wogan GO, Swartz B, Quental TB, Marshall C, McGuire JL, Lindsey EL, Maguire KC and Mersey B (2011) Has the Earth's sixth mass extinction already arrived? *Nature* 471(7336): 51.

Bartel R, Graham N, Jackson S, Prior JH, Robinson DF, Sherval M and Williams S (2013) Legal geography: An Australian perspective. *Geographical Research* 51(4): 339–353.

Bennett NJ, Roth R, Klain SC, Chan K, Christie P, Clark DA, Cullman G, Curran D, Durbin TJ, Epstein G and Greenberg A (2017) Conservation social science: Understanding and integrating human dimensions to improve conservation. *Biological Conservation* 205: 93–108.

Blicharska M, Orlikowska EH, Roberge JM and Grodzinska-Jurczak M (2016) Contribution of social science to large scale biodiversity conservation: A review of research about the Natura 2000 network. *Biological Conservation* 199: 110–122.

Brechin SR, Wilshusen PR, Fortwangler CL and West PC (2002) Beyond the square wheel: Toward a more comprehensive understanding of biodiversity conservation as social and political process. *Society and Natural Resources* 15: 41–64.

Chan KM, Pringle RM, Ranganathan JAI, Boggs CL, Chan YL, Ehrlich PR, Haff PK, Heller NE, Al-khafaji K and Macmynowski DP (2007) When agendas collide: Human welfare and biological conservation. *Conservation Biology* 21(1): 59–68.

Costanza R, De Groot R, Sutton P, Van der Ploeg S, Anderson SJ, Kubiszewski I, Farber S and Turner RK (2014) Changes in the global value of ecosystem services. *Global Environmental Change* 26: 152–158.

Di Marco M, Ferrier S, Harwood TD, Hoskins AJ and Watson JE (2019) Wilderness areas halve the extinction risk of terrestrial biodiversity. *Nature* 573(7775): 582–585.

Gillespie J (2018) Wetland conservation and legal layering: Managing Cambodia's great lake. *Geographical Journal* 184(1), 31–40.

IPBES (Intergovernmental Science-Policy Platform on Biodiversity and Ecosystem Services) (2019) *Global Assessment Report*. Available https://ipbes.net/global-assessment. Accessed 1 October 2019.

IUCN (International Union for the Conservation of Nature) and World Commission on Protected Areas (WCPA) (2017) *IUCN Green List of Protected and Conserved Areas: Standard Version 1.1*. IUCN, Gland, Switzerland. Available www.iucn.org/greenlist. Accessed 13 June 2019.

Johnson CN, Balmford A, Brook BW, Buettel JC, Galetti M, Guangchun L and Wilmshurst JM (2017) Biodiversity losses and conservation responses in the Anthropocene. *Science* 356(6335): 270–275.

Jones KR, Venter O, Fuller RA, Allan JR, Maxwell SL, Negret PJ, and Watson JE (2018) One-third of global protected land is under intense human pressure. *Science* 360(6390): 788–791.

Joppa LN, Bailie JE and Robinson JG (2016) *Protected Areas: Are They Safeguarding Biodiversity?* Wiley Blackwell and Wiley, UK.

Lausche BJ and Burhenne-Guilmin F (2011) IUCN Guidelines for protected areas legislation (No. 81), IUCN, Gland, Switzerland. Available https://data.iucn.org/dbtw-wpd/edocs/EPLP-081.pdf. Accessed 28 February 2019.

Mappin B, Chauvenet AL, Adams VM, Di Marco M, Beyer HL, Venter O, Halpern BS, Possingham HP and Watson JE (2019) Restoration priorities to achieve the global protected area target. *Conservation Letters*: 12646.

Mascia MB, Brosius JP, Dobson TA, Forbes BC, Horowitz L, McKean MA and Turner NJ (2003) Conservation and the social sciences. *Conservation Biology* 17(3): 649–650.

Massey D (2005) *For Space*. Sage, London.

Mora C, Tittensor DP, Adl S, Simpson AG and Worm B (2011) How many species are there on Earth and in the ocean? *PLoS Biology* 9(8): 1001127.

O'Donnell T (2020) Inside-outside: An interrogation of coastal climate change adaptation through the gaze of 'the lawyer'. In: O'Donnell T, Robinson DF and Gillespie J (eds) *Legal Geography: Perspectives and Methods*. Routledge, London.

O'Donnell T, Robinson DF and Gillespie J (eds) (2020) *Legal Geography: Perspectives and Methods*. Routledge, London and New York.

Olwig K (1995) Reinventing common nature: Yosemite and Mt. Rushmore—A meandering tale of a double nature. In: Cronan W (ed) *Uncommon Ground: Toward Reinventing Nature*. W. W. Norton, pp. 379–408.

Orzeck R and Hae L (2019) Restructuring legal geography. *Progress in Human Geography*. Available https://doi.org/10.1177/0309132519848637.

Pimm SL, Jenkins CN and Li BV (2018) How to protect half of Earth to ensure it protects sufficient biodiversity. *Science Advances* 4(8): 2616.

Sala OE, Chapin FS III, Armesto JJ, Berlow E, Bloomfield J, Dirzo R, Huber-Sanwald E, Huenneke LF, Jackson RB, Kinzig A, Leemans R, Lodge DM, Mooney HA, Oesterheld M, Poff NL, Sykes MT, Walker BH, Walker M and Wall DH (2000) Global biodiversity scenarios for the year 2100. *Science* 287: 1770–1774.

Schaaf T and Clamote Rodrigues D (2016) Managing MIDAs: Harmonising the management of multi-internationally designated areas: Ramsar Sites, World Heritage sites, Biosphere Reserves and UNESCO Global Geoparks. IUCN, Gland, Switzerland.

Steffen W, Richardson K, Rockström J, Cornell SE, Fetzer I, Bennett EM, Biggs R, Carpenter SR, De Vries W, De Wit CA and Folke C (2015) Planetary boundaries: Guiding human development on a changing planet. *Science* 347(6223): 1259855.

Treves A, Wallace RB, Naughton-Treves L and Morales A (2006) Co-managing human–wildlife conflicts: A review. *Human Dimensions of Wildlife* 11(6): 383–396.

UNEP (United Nations Environment Program) and WCMC (World Conservation Monitoring Centre) (undated) *World Database on Protected Areas*. Available https://www.protectedplanet.net/. Accessed 5 October 2019.

Watson JE, Jones KR, Fuller RA, Marco MD, Segan DB, Butchart SH, Allan JR, McDonald-Madden E and Venter O (2016a) Persistent disparities between recent rates of habitat conversion and protection and implications for future global conservation targets. *Conservation Letters* 9(6): 413–421.

Watson JE, Shanahan DF, Di Marco M, Allan J, Laurance WF, Sanderson EW, Mackey B and Venter O (2016b) Catastrophic declines in wilderness areas undermine global environment targets. *Current Biology* 26(21): 2929–2934.

Watson JE, Venter O, Lee J, Jones KR, Robinson JG, Possingham HP and Allan JR (2018) Protect the last of the wild. *Nature* 563: 27–30.

Whitttaker RJ, Araujo MB, Jepson P, Ladle RJ, Watson JEM and Willis KJ (2005) Conservation biogeography: Assessment and prospect. *Diversity and Distributions: A Journal of Conservation Biogeography* 11(1): 3–23.

WWF (World Wildlife Fund) (2018) *Living Planet Report*. Available https://www.zsl.org/global-biodiversity-monitoring/indicators-and-assessments-unit/living-planet-index. Accessed 3 October 2019.

CHAPTER 2

Protected Areas

Abstract This chapter considers the phenomenon of in situ conservation efforts through protected areas. The chapter proceeds in three parts. The first part defines protected areas and outlines the (relatively) recent history of the idea. The protected area movement is very much a phenomenon born from the nineteenth century romanticisation of landscape, most famously by luminaries such as Wordsworth, Muir and many others. Today, it is the major policy tool for environmental conservation. The second section in this chapter considers *why* protected areas are an essential part of the modern conservation agenda. The problem of rapidly accelerating biodiversity decline; devastating losses of habitats, ecosystems and species keeps in situ conservation at the forefront of efforts to halt or stem systemic collapse. The final part of this chapter looks at the failures of protected areas and argues that an approach informed by a different way of thinking is needed. Accordingly, we turn to consider how a legal geography lens could improve current governance and management challenges in protected areas.

Keywords Protected areas · In situ conservation · National park · IUCN · Aichi Targets · Legal geography

J. Gillespie, *Protected Areas*,
https://doi.org/10.1007/978-3-030-40502-1_2

2.1 Introduction

I spent a considerable part of my childhood exploring the Australian countryside and bush, none of it untouched, but all mostly at a distance from urban landscapes. It seems, thanks to the vicissitudes of memory, an idyllic scene. Many hours, it seems months and years, were spent scrambling across steep, tea-tree covered hillsides, traipsing on my pony across paddocks, jumping creeks, crashing through undergrowth and meandering through the dusty dirt roads. Of course, it was not such simple a thing, with much toil and labour and the long-forgotten frustration of rural isolation. However, one element of such a childhood continues to capture my imagination—an appreciation of the world around me. Observation, description and appreciation of the world in which we live, in whatever form it takes, holds my attention. An understanding that we are part of this scene, not removed from it, and that this scene and our involvement in it is worth documenting and protecting is a sentiment that underpins the modern protected area approach to conservation. The words of John Muir (1911, p. 10), noted naturalist, conservation and founder of the Sierra Club, are evoked:

> We are now approaching the region of clouds and cool streams. Magnificent white cumuli appeared about noon above the Yosemite region … No rock landscape is more varied in sculpture, none more delicately modelled than these landscapes of the sky; domes and peaks rising swelling, white as finest marble and firmly outlined, a most impressive manifestation of world building. Every rain-cloud, however fleeting, leaves its mark, not only on trees and flowers whose pulses are quickened, and on the replenished streams and lakes, but also on the rocks are its marked engraved whether we see them or not.

Muir's words, although written in 1869 in a vastly different place, resonate today. John Muir is considered, after all, to be the 'founding father' of national parks.

2.2 Understanding Protected Areas

Designation of a protected area requires that a particular location is defended or shielded against harms—generally by restricting human activities through law—because the location has environmental or cultural value. While the practice of enacting protected areas differs depending on the national, regional or local setting, the unpinning concept of protecting through the management (and often exclusion) of human actions permeates all forms of protective operations. Protected area systems are the foundations of conservation strategies across the globe.

The most widely used and widely accepted definition of a protected area comes from the work of the International Union for the Conservation of Nature (IUCN) which defines a protected area as;

> A clearly defined geographical space, recognised, dedicated and managed, through legal or other effective means, to achieve the long-term conservation of nature with associated ecosystem services and cultural values. (IUCN undated)

The IUCN was created in 1948 as a global conservation organisation composed of voluntary experts (the 'union') from government, academia (such as I, a member of the union) and civil society who provide expertise on the conservation of the natural world (IUCN undated). The IUCN has defined six categories of protected area according to their management objectives. These categories are regarded as setting the global standard for defining protected areas and often form the basis for, or inform, national legislation for protected areas. The categories range from Ia Strict Nature Reserves and Ib Wilderness Areas to II National Parks, III National Monuments or Features, IV Habitat Species Management Areas, V Protected Landscape/Seascape and VI Protected areas with sustainable use of natural resources (IUCN undated).

In July 2018 it was estimated that protected areas cover approximately 20 million km^2 of the earth's land surface and approximately 27 million km^2 of its oceans (UNEP-WCMC, IUCN and NGS 2018). This represents 238,563 designated protected areas comprising of 14.9 and 7.3% of the land and oceans across the globe (UNEP-WCMC, IUCN and NGS 2018, p. 6). It is clear that protected areas are now the primary regulatory and policy tool for environmental conservation, but how did this global in situ conservation juggernaut—the protected area—emerge? We turn to consider the emergence of this important global environmental policy phenomenon (Chape et al. 2008), despite the oft-cited warning that tracing the origins of the protected area concept, through the phenomenon of the national park, is 'like nailing jelly to the wall' (Everhart 1983).

The colonialization of the New World took place in an era of exploration, especially in the occupation of Australasia and the Americas (Sheail

2010). The ill-effects of colonialization across the globe remain with us to the present, and a growing de-colonization message forces us to carefully re-consider the intergenerational impacts of contentious colonialization processes. The modern protected area movement is a legacy of this era, but does not necessarily draw as much critical attention as we might expect in our decolonialization agendum. Traced through iconic sites such as Yellowstone and Yosemite to the Royal National Park in Australia, formal protected areas, endorsed through state and federal government regulation, form a key component of biodiversity conservation programmes across the world. We will turn to the biodiversity protection component of protected areas later in this chapter.

Before considering the New World experience, I want to sound a note of caution. There have been many scholars, activists, consultants and beyond, assessing and critiquing protected area policy and programs for many decades. Within the academy notable commentators are working at the social sciences/conservation nexus in and around this issue (see, for some examples, Brockington 2002; Dowie 2009; Adams and Hutton 2007; Adams et al. 2014; Adams 2019). Their work draws our attention to the importance of placing people in the conservation picture, and, especially, of the distorting effects of ignoring people (see, for example, work in the 1990s by Adams and McShane 1996). Dowie (2009) articulates the problematics of simplifying the history of the conservation movement vis-à-vis indigenous communities where active dispossession through conservation continues to suppress indigenous voices and interests across the globe. Displacement, relocation, forced migration—all have a hand to play in how the modern protected area network has come into being. Exclusion of people from place sits squarely within the conservation objective epitomised through the protected area movement. West and Brockington (2006, p. 610) put this succinctly,

> (P)rotected areas necessarily seek to protect nature and biodiversity by abstracting them from their complex social contexts. Much protected-area legislation and policy is a practical exercise in virtualism—defining what human behavior and actions ought to be in a particular location and restricting it accordingly.

In a similar vein, Cronon (1995, p. 79) provides a clear insight in describing the exclusion of traditional owners from 'pristine' wilderness during the establishment of protected areas in the United States,

> The movement to set aside national parks and wilderness areas followed hard on the heels of the final Indian wars, in which the prior human inhabitants of these areas were rounded up and moved onto reservations. ...original inhabitants were kept out by dint of force, their earlier uses of the land redefined as inappropriate or even illegal.

In the process of setting aside protected areas we bound and exclude, meaning past inhabitants and users are also excluded (Dowie 2009). In recent decades there has been a concerted effort to engage traditional owners through community-based management in a 'wise use' paradigm. Engagement with customary owners is seen as an important part of the decolonisation process and empowerment of displaced or otherwise marginalised peoples (Ross et al. 2009; De Lacy 1994). Bringing such co-management arrangements in line with the overarching need to maintain the ecological status of protected areas remains, as we shall see, a significant challenge.

To contextualise the protected area movement, we need to consider how the more modern iteration of this experience began. The protected area viewed through the 'national park' ideal is a significant development in the history of conservation (Phillips 2004). From the 1800s in New World countries especially, we can trace a movement to reserve parcels of land which sets the scene for the global protected area juggernaut that exists today (Crombie 2014). Claims for 'first' national park status are debated. Gillespie (2007, p. 8) suggests that the United States' Congress presentation of part of today's Yosemite National Park's to the state of California in 1864 marks the start of the formalised national park movement. However, the notion of setting aside tracts of land or water for preservation and/or conservation pre-dates the New World phenomenon by millennia, and others have done more justice to that wider historical setting than I propose to do here (see, for example, Gissibl et al. 2012).

In a great rush during period from the 1860s through the 1870s, the world witnessed the establishment of Yellowstone (1872), Royal National Park (1866), Canadian Rockies (1879) and Tongariro in New Zealand (1877). In 1898 South Africa dedicated the Krueger National Park (Gillespie 2007). However, for many, this story begins with the creation of Yosemite National Park and its official designation through federal (US) law in 1864 (Sheail 2010).

Yosemite was designated as a State park by a special Act of Congress (1864), thus creating a mandated protected area for the Yosemite Valley situated within the state territory of California (Ogden 2015). The surrounding Sierra Nevada mountains became part of three new national (i.e. federally protected) parks: Yosemite, Sequoia and Kings Canyon, in 1890 (Ogden 2015). Yosemite's protected status sat in contrast to the urban parks established up to that time in the United States (e.g. Boston Common, Massachusetts, 1634), which were crafted, managed land and waterscapes. President Lincoln signed the bill establishing the Yosemite Grant on 30 June 1864. The Californian state legislature recognised the park in 1866 (Ogden 2015). Law gave this place its status; without the codification to set aside this place it would not exist in the manner we see it today. Law ensures that this landscape of grassy meadows, winding rivers, managed orchards (from early settlers) and rugged mountain ranges (Ogden 2015) is 'frozen' for all to enjoy, in perpetuity. The use of formal laws to set aside land and waterscapes in the United States was cemented with the establishment of Yellowstone National Park in 1872, widely described as the world's first national park.

But the history of codification for (national) parks as protected places is not, perhaps, all as it seems at first glance (Robin 2013). As Robin explains, Yellowstone was not the world's first national park, but it is celebrated as such for three reasons: first, the timing of global centennial celebrations in the 1970s coincided better with Yellowstone's legal designation (Yosemite's protected status a decade earlier did not coincide with the movement/mood of the 1970s); second, Yellowstone was truly a national park, having been created with federal legislation across state boundaries (technically Yosemite was within one state, California) and, finally, Yellowstone's important symbolic value as a cultural iconic place associated with the American West. Technically speaking, Yosemite was initially dedicated as a state park, and it did not become a national park until 1890. Moreover, it did not fall under US federal jurisdiction until 1906 (Crombie 2014).

One of the reasons we focus on the evolution of protected areas in the post-New World era is due to the circumstances that characterise nation building in this era. It is often said that the national park was "America's invention" (Nash 1970). Nash's descriptor is based on his observation that the United States would not lead the championing of protected areas (through national parks) but for four factors: (1) The experience and valorisation of the American 'wilderness' ideal, (2) The democratic state allowing land to be held in trust by the state (government/s) for the people (3) The significant 'undeveloped' (notwithstanding pre-existing populations) lands and (4) The affluence of the nation such that land could be set aside without the need to convert it for cultivation/development. Other New World nations shared these attributes.

Developments surrounding the establishment of formal protected areas were also afoot in the Australasian region. In 1866 the Royal National Park, south of Sydney in the state of New South Wales, was created. Again, however, this was not technically a national park in the sense that it was not created by national or federal legislation, simply because federation would not occur until the Commonwealth of Australia came into being in 1901. Nonetheless, it was described as a 'National' park, being set aside as a reserve on existing public (Crown) lands. Reserving public lands in this way was novel (Crombie 2014). The creation of the Royal National Park in New South Wales is generally credited or at least associated with Sir John Robertson, a New South Wales parliamentarian (Scott 2011). The motivations for its establishment are analogous to those that led to the foundation of Yellowstone in the United States, in that these areas were set aside for the people—for social and recreational purposes—but in Australia it seems that the idea of any underpinning motive to protect natural attributes, or any underlying conservation ethic, was not as clear. Scott (2011) argues, however, that an emergent conservation ethic is evident in the appointment of rangers to the park to prevent the removal of plants and to prevent destruction of the natural landscape.

The presence, or absence, of an explicit conservation ethic that underpinned the creation of formal protected areas was not decisive in their success or popularity. 'Conservation thinking' as we might recognise it today, was not necessarily pivotal in the early development of the national park movement but has certainly become so over time and is now the primary justification for the establishment and maintenance of protected areas. The desire to valorise, romanticise or simply understand the symbiotic relationship between 'nature' or 'wilderness' and humans is evident in classical texts that have come to define the founding principles of conservation, including Marsh (1864), Thoreau (1854), Muir (1911), Leopold (1949) and others. Preserving and conserving places for 'nature' demands that we think about what we mean when we use that word. Increasingly, protected areas are justified in terms of the maintenance of ecosystem services upon which humans rely, either directly or indirectly, and this is often expressed as a need to conserve the diversity and abundance of species.

2.3 The Biodiversity Imperative

Prevention of stark and increasing biodiversity losses (Díaz et al. 2019) takes place through improving ecosystem and habitat protection, and this directly implicates the critical importance of protected area goals. Within the global environmental framework of the *Convention on Biological Diversity*, a binding agreement borne out of the 1992 Rio Earth Summit, nation states have created international biodiversity protection obligations. This agreement came into force in 1993 and the *Convention* has (almost) universal commitment (with the startling exception of the USA). The 2011–2020 Strategic Plan for the *Convention* set 20 targets—known as the Aichi Targets—to meet the overarching goal: to improve the status of biodiversity by safeguarding ecosystems, species and genetic diversity (Goal C). Aichi Target 11 is of particular relevance for it articulates an ambition to enhance the global coverage of protected areas:

> By 2020, at least 17 per cent of terrestrial and inland water areas and 10 per cent of coastal and marine areas, especially areas of particular importance for biodiversity and ecosystem services, are conserved through effectively and equitably managed, ecologically representative and well-connected systems of protected areas and other effective area-based conservation measures, and integrated into the wider landscape and seascape.

The rationale for this is:

> Well governed and effectively managed protected areas are a proven method for safeguarding both habitats and populations of species and for delivering importance ecosystem services. (see CBD undated)

There is absolutely no doubt—at the international level a traditional protected area approach is *the* vital policy response for protecting habitat and ecosystem function for planetary health. Yet, despite the dedicated initiatives to stem the degradation of habitat, decline continues with widespread and aggressive ecosystem losses globally (for example, see Watson et al. 2018). In these circumstances, inevitably, we must question whether global conservation policy is doing enough to slow losses and question whether the protected area approach is succeeding as well as it might.

A key component of IUCN work involves defining and re-defining protected areas. In 2008 the IUCN slightly changed the definition to ensure that the conservation of nature was the primary/priority aim for protected areas (Lopoukhine and de Souza Dias 2012). In 2010 the definition was expanded to include 'other effective area-based conservation measures' ('OECMs'). Jonas et al. (2014) called for guidance on what OECMs would mean for meeting conservation targets. Laffoley et al. (2017, p. 131) explain that the addition of the OECMs was a "last-minute addition to acknowledge the contribution that many areas not legally designated as protected areas can contribute to effective conservation." They also note that the primary method to achieve Target 11 would remain formally designated protected areas. While Lopoukhine and de Souza Dias (2012) argue that care is needed in framing approaches based on OECMs, for some opening of the idea of protected areas to (e.g. intensive forestry) 'would seriously undermine their biodiversity values' (p. 6) and could, in fact, produce perverse outcomes. While caution is needed in this respect, as Lopoukhine and de Souza Dias identify, there is also no doubt that widespread biodiversity conservation requires efforts beyond protected area borders.

While this book advocates for the continuation of the protected area ideal, it does not do so naively or without critique. Integrity in landscapes beyond the boundaries of protected areas is also critical for species conservation. Habitat fragmentation causing disjunction in species distributions need not necessarily be problematic, but in times of rapid environmental change habitat fragmentation can make concurrent biodiversity losses worse. Striving towards enhanced ecological resilience is both intrinsically (from an ecocentric perspective) and selfishly (from an anthropocentric perspective) valuable. This is not unrelated to a call to improve existing protected area management approaches, as Worboys et al. (2010, p. 7) observe,

> (C)onnectivity conservation areas are not a substitute for protected areas. ... However, it will not be possible to achieve reserve establishment for all natural lands and this is why connectivity areas are an important next conservation initiative for nations.

This does not, as they say, replace protected areas—for the purposes of connectivity protected areas continue to play a vital role in biodiversity conservation. The author's note, however, a key concern that management practices for protected areas require quality planning, sufficient resources and on-the-ground implementation in order for associated connectivity efforts to succeed (Worboys et al. 2010, p. 344).

2.4 Challenges and Next Steps: The Value of a Legal Geography Approach

If current approaches to conservation are both ubiquitous and failing, we must consider whether different approaches might help forge new directions and new priorities in protected area governance that, ultimately, improve conservation outcomes for all those communities (human and non-human) that constitute these special places.

Woodley et al. (2012) comment that, to achieve Target 11 of the Aichi Targets of the *Convention on Biological Diversity* Strategic Plan (2011–2020) for protected areas (in safeguarding ecosystems by conserving habitat), multiple dimensions need to be addressed, namely: (1) increasing spatial coverage, (2) increasing connectivity, (3) ensuring effective management/governance and (4) achieving equitable outcomes. As Woodley et al. (2012) note:

> Protected area planning and management is a complex field requiring countries to consider spatial planning, ecological connectivity, and integration with surrounding landscapes, as well as effective management and a range of social issues around equity. Target 11 is remarkably dense in its construction and implication. It contains all the necessary elements for success, but countries need to consider all the elements as a whole to be successful. For many countries the pieces of this protected area puzzle are broken or do not exist. The pieces will have to be built or repaired in order to construct the whole. (p. 33)

It is in this context that concern with the management and governance dimensions of protected area objectives come to the fore. The contextualised social-ecological-regulatory issues associated with particular places that support or undermine effective management approaches is elaborated further in later chapters. The situation as it presently stands is that there is slow progress on the 'people' dimensions of the Targets (associated with improving governance regimes). A recalibrated focus on the substance (in laws) and processes (in laws) and how these link to practices by capturing data about the everyday lived experiences of those impacted by protected areas is urgently needed. Recalling the work of (for example) Cronon

(1995) and Nash (1970), we need to put the people being regulated back into this protected area picture; we must not abstract people/place connections in crafting regulatory solutions in our pursuit of habitat/species conservation. To do so invites conflict, lack of compliance and regulatory failure. Best-practice requires that we understand people/place connections before we enact any measures to protect or exclude, and in doing so we are performing a legal geography analysis.

The rationale for this book lies in linking protected area management to legal geography approaches that put the law-people-place dynamics at the centre of this in situ conservation agenda. The conceptual and applied repertoire of legal geography (see O'Donnell et al. 2020), is essential in understanding the creation, monitoring and on-going management of protected areas. Gissibl et al. (2012) also argue that the emergence of institutions to support the growing park movement in the global realm (e.g. IUCN) created homogenous science-based approaches to environmental policy. But this institutional approach has its shortcomings;

> Parks are treated as part and parcel of an emerging world environmental regime that is largely Western in its origin, and there is little sensitivity for the political asymmetries, conflicting interactions and the contestations surrounding the adaptation of the park concept to diverse political and cultural settings. (p. 6)

An holistic perspective, encouraged by legal geographers, can reframe thinking about protected areas in useful ways. As I have argued, importing the (Western) protected area idea—through international law—into and onto 'other' places is inherently problematic (see also Grebowicz 2015). We need to interrogate the roll-out of protected areas across the globe and be wary of their homogenising effect. We need to directly engage with agency in this exercise. It is through the work of legal geographers—who look at law, lore, people and places simultaneously and not in isolation—that we can expose the assumptions embedded in the protected area ideal. In this sense, legal geography scholarship directly engages with agency (in direct contrast to the position suggested by Orzeck and Hae 2019; see also O'Donnell et al. 2020).

So, how can we do better? One way is to carefully link places (habitats) to people before we write the laws which restrict what humans can and cannot do. This is a call to arms in the face of spiralling habitat loss: follow the best science but articulate this with perspectives drawn from human communities living, working, and visiting these special places. This goes beyond a call for more (or better) social science in protected area policy development, because the law component has been (largely) overlooked in existing social science work relating to protected areas. While the focus on anthropology, sociology, economics and political dimensions are important, we should not overlook the law-social-enviro dimensions of the protected area policy arena.

Bringing the legal-socio-enviro together can, in fact, provide a much needed refocus (Benidickson and Paterson 2017) to protected area policy evolution and refinement. This focus complements the IUCN protected area standards with its emphasis on good governance, sound planning and design, effective management and successful conservation outcomes IUCN and World Commission on Protected Areas (WCPA) (2017). This is because there has been a weakness in the translation phase.

In many cases, attention to place and the lived experiences of the people impacted by protected areas is displaced to a monitoring or reporting phase of the protected area process. Poorly crafted regulations that abstract the nuance of local conditions may be significantly out-of-sync with community expectations and thus create tension and invite non-compliance, in turn requiring enforcement/penalty. Ultimately, this regulatory failure *enables* environmental degradation and habitat fragmentation, leading to further biodiversity losses. The contribution legal geography can make to the protected area process, by paying close attention to the place/people/law dynamic, is to facilitate regulation that draws community expectations and conservation science/policy more closely together, avoiding or mitigating regulatory failure. We turn to investigate legal geography further in the next chapter—what is it and why can this way of thinking can aid/assist the protected area conservation agenda.

References

Adams JS and McShane TO (1996) *The Myth of Wild Africa: Conservation Without Illusion*. University of California Press, Berkeley.

Adams WM (2019) Geographies of conservation III: Nature's spaces. *Progress in Human Geography*. Available https://doi.org/10.1177/0309132519837779.

Adams WM and Hutton J (2007) People, parks and poverty: Political ecology and biodiversity conservation. *Conservation and Society* 5(2): 147.

Adams WM, Hodge ID and Sandbroo L (2014) New spaces for nature: The re-territorialization of biodiversity conservation under neoliberalism in the UK. *Transactions of the Institute of British Geographers* 39: 574–588.

Benidickson J and Paterson A (2017) Biodiversity, protected areas and the law. In: McManis CR and Ong B (eds) *Routledge Handbook of Biodiversity and the Law*. Routledge, London, pp. 42–59.

Brockington D (2002) *Fortress Conservation: The Preservation of the Mkomazi Game Reserve, Tanzania*. Indiana University Press, Bloomington.

CBD (Convention on Biological Diversity) (undated) *Target 11 Technical Rationale Extended*. Available https://www.cbd.int/sp/targets/rationale/target-11/. Accessed 15 September 2019.

Chape S, Spalding M and Jenkins M (2008) *The World's Protected Areas: Status, Values and Prospects in the 21st Century*. University of California Press, California.

Crombie B (2014) *Which was the world's first national park*. National Parks Association of New South Wales. Available https://npansw.org/2017/09/01/which-was-the-worlds-first-national-park/. Accessed 30 November 2019.

Cronon W (1995) The trouble with wilderness; or getting back to the wrong nature. In: Cronon W (ed) *Uncommon Ground: Rethinking the Human Place in Nature*. W. W. Norton, New York, pp. 69–90.

De Lacy T (1994) The ulu ru/kakadu model—an angu tjukurrpa. 50,000 years of aboriginal law and land management changing the concept of national parks in Australia. *Society and Natural Resources* 7(5): 479–498.

Díaz S, Settele J, Brondizio ES, Ngo, HT, Guèze M, Agard J, Arneth A, Balvanera P, Brauman K, Butchart S and Chan K (2019) *Summary for policymakers of the global assessment report on biodiversity and ecosystem services of the Intergovernmental Science-Policy Platform on Biodiversity and Ecosystem Services*. IPBES Secretariat, Bonn, Germany.

Dowie M (2009) *Conservation Refugees: The Hundred Year Conflict Between Global Conservation and Native Peoples*. MIT Press, Cambridge, MA.

Everhart W (1983) *The National Park Service*. Routledge, New York and Abingdon.

Gillespie A (2007) *Protected Areas and International Environmental Law*. BRILL, The Netherlands.

Gissibl B, Hohler S and Kupper P (2012) *Civilizing Nature: National Parks in Global Historical Perspective*. Berghahn Books, Oxford.

Grebowicz M (2015) *The National Park to Come*. Stanford University Press, Stanford.

IUCN (International Union for the Conservation of Nature) (undated) *About*. Available https://www.iucn.org/about. Accessed 15 September 2019.

IUCN (International Union for the Conservation of Nature) and World Commission on Protected Areas (WCPA) (2017) *IUCN Green List of Protected and*

Conserved Areas: Standard Version 1.1. IUCN, Gland, Switzerland. Available www.iucn.org/greenlist. Accessed 13 June 2019.

Jonas H, Barbuto V, Jonas HC, Kothari A and Nelson F (2014) New steps of change: Looking beyond protected areas to consider other effective area based conservation measures. *Parks* 20(2): 111–128.

Laffoley D, Dudley N, Jonas H, MacKinnon D, MacKinnon K, Hockings M and Woodley S (2017) An introduction to 'other effective area-based conservation measures' under Aichi Target 11 of the Convention on Biological Diversity: Origin, interpretation and emerging ocean issues. *Aquatic Conservation: Marine Freshwater Ecosystems* 27(S1): 130–137.

Leopold A (1949) *A Sand County Almanac and Sketches Here and There.* Oxford University Press, Oxford and New York.

Lopoukhine N and de Souza Dias BF (2012) What does Target 11 really mean? *Parks* 18(1): 5.

Marsh GP (1864) *Man and Nature*, reprint 2003, University of Washington Press, Seattle and London.

Muir J (1911) *My First Summer in the Sierrra*, reprint 2004, Dover Publications Inc., Mineola, NY.

Nash R (1970) The American invention of national parks. *American Quarterly* 22: 726–735.

O'Donnell T, Robinson DF and Gillespie J (eds) (2020) *Legal Geography: Perspectives and Methods.* Routledge, London and New York.

Ogden KN (2015) *Yosemite.* Reaktion Books Limited, London.

Orzeck R and Hae L (2019) Restructuring legal geography. *Progress in Human Geography.* Available https://doi.org/10.1177/0309132519848637.

Phillips A (2004) The history of the international system of protected area management categories. *Parks* 14(3): 4–14.

Robin L (2013) Being first: Why the Americans needed it, and why Royal National Park didn't stand in their way. *Australian Zoologist* 36(3): 321–331.

Ross H, Grant C, Robinson CJ, Izurieta A, Smyth D and Rist P (2009) Co-management and indigenous protected areas in Australia: Achievements and ways forward. *Australasian Journal of Environmental Management* 16(4): 242–252.

Scott KA (2011) Shared ideals in Yellowstone and Royal National Parks. *Yellowstone Science* 19(3): 6–11.

Sheail J (2010) *Nature's Spectacle: The World's First National Parks and Protected Places.* Earthscan, London.

Thoreau HD (1854) *Walden.* Oxford University Press, London.

UNEP-WCMC, IUCN and NGS (2018) *Protected Planet Report 2018.* UNEP-WCMC, Cambridge, UK; IUCN, Gland, Switzerland; and NGS, Washington, DC, USA.

Watson JE, Venter O, Lee J, Jones KR, Robinson JG, Possingham HP and Allan JR (2018) Protect the last of the wild. *Nature* 563: 27–30.
West P and Brockington D (2006) An anthropological perspective on some unexpected consequences of protected areas. *Conservation Biology* 20(3): 609–616.
Woodley S, Bertzky B, Crawhall N, Dudley N, Londoño JM, MacKinnon K, Redford K and Sandwith T (2012) Meeting Aichi Target 11: What does success look like for protected area systems. *Parks* 18(1): 23–36.
Worboys G, Francis WL and Lockwood M (2010) *Connectivity Conservation Management: A Global Guide*. Earthscan, London.

CHAPTER 3

A Legal Geography Approach

Abstract This chapter articulates the benefits of a legal geography approach for a descriptive, reflective and, ultimately, a critical analysis of protected areas. I argue that legal geography scholarship has come-of-age; it is a perspective for our times. In a world shaped increasingly by law, and troubled by complex multi-causal environmental challenges, an ability to understand the nexus between regulation, people and place is an essential skill for multidisciplinary successes in both research and policy. Legal geographers must learn data analysis methods from within the law and geography disciplines; they can read maps, interpret biophysical landscapes, or use social science methods such as interviewing to understand how and what people think, while reading and understanding parliamentary debates and case law. These skills are inherently valuable in our fast-changing world.

Keywords Legal geography · Place · Property · Territory · Political ecology · Environmental justice · Ecological justice · Legal ecology · Global South · More-than-human

3.1 Introduction

In this chapter, I encourage us to consider the relevance of contemporary legal geography work to demonstrate how this perspective can aid our thinking about protected area management. There are six sections in this

J. Gillespie, *Protected Areas*,
https://doi.org/10.1007/978-3-030-40502-1_3

chapter. The first section encourages us to consider the world around us within geographical and regulatory frames. From this we can visualise a complexity that, in turn, encourages us to recognise and critique the ways we shape and re-shape our lived-in worlds. The second section describes the evolution of this exciting and important new direction in scholarship and practice. The third section delves into the importance of place-based research that is focussed on the material, lived-in land- and waterscapes. Beyond these framings, there are important component parts of a legal geography approach that identify the problematic dynamics of property and territory in place-making that often give rise to conflict and environmental degradation. The linkages between these areas of scholarship are explored in fourth section. In fifth section we turn to consider the conceptual nuances of today's legal geography research. Legal geographers, through their scrutiny of regulatory process, begin to forge links with dominant conceptual themes from geography, especially in Political Ecology and Environmental/Ecological Justice. In the final section of this chapter we explore how the work of legal geographers is becoming crucial in unpacking the complexities of enviro-social-legal meaning, especially in the Global South and in the context of a wider societal decolonisation project where recognition and engagement is increasingly in demand. In particular, legal geographical scholarship can explain the dynamic nature of institutional arrangements and, critically, can identify ambiguity in these relationships. Legal geography research can also take us, critically, into the more-than-human world to give voice to those that are suppressed or silenced in regulatory decision-making processes.

3.2 Setting a Scene: The Importance of Reading and Engaging Place

On a blisteringly cold winter's morning on the edge of the Derwent River in Hobart, Tasmania, Australia, I sit in the comfort of a heated room and, through a bay window, consider a wonderful panorama of the river. From where I sit, I can see the road that runs along the foreshore, the footpath, the boat moorings, rowers and a public park. I silently watch a passing pedestrian, car and boat traffic come and go in their daily rituals of events and commutes. Not all are confined to the usual daily trudge (or sprightly walk—depending on your daily mood) to and from the office, however. These rituals include toddlers playing on the swings, a temporary but obsessive pastime, and the daily dog walk with fresh-faced puppies playfully circling old, weary and aloof dogs who have long-ago marked out this place as their own. I reflect on how I read this scene—this land and waterscape—with new and old patterns repeating over and over in place and time.

I find myself looking for insignia to read the movement before me. The playground is fenced with a lockable gate. The sign prohibits dogs from entering this space. Outside of the playground, off-leash times for the park are signposted clearly. Labels, both real and fixed and imagined and learned, guide the movements of every single human, and some non-humans, in this scene. We are all trained to understand how to read this land and waterscape. To know what behaviours are and are not acceptable to our fellow humans. Underpinning our actions are two essential components—the place and the law. The biophysical traits of place, a cold Derwent River, on crisp, sunny winter morning prevent us from entering the water for a swim, for example. Place makes its impact on our behaviour. The signs, both explicit and implicit, regulating our use of the road, footpath, park, and waterway give us the legal overlay that controls our behaviours. Together, place and law are entwined to produce an encounter that we understand in the context of this scene. We can read our land and waterscape through the twin messages provided through the lens of place and law, in what we now describe as legal geography. It seems odd that law has not engaged more readily with the ideas ingrained in geography surrounding how human-environment interactions emerge, for, as Nagle (2010, p. 4) observes it is crucial that environmental law engages with place:

> (O)ne would expect that a legal regime premised upon the importance of preserving or recovering certain conditions within the natural environment would pay more attention to what those conditions actually are ... a body of law designed to incorporate holistic ecological principles should recognize the importance of a holistic examination of all of the laws that affect the natural environment, but too often that has not been the case.

A comprehensive, integrated and holistic perspective on the human-environment-regulatory nexus is essential to effectively dealing with accelerating environmental changes.

3.3 Understanding Legal Geography's Importance

Legal geography scholarship has created a growing and increasingly persuasive body of literature that exposes and analyses critically the place-people-law nexus. David Delaney, in three recent review papers (Delaney 2015, 2016, 2017), has neatly captured the evolution of the legal geography project. In describing the evolution of the field, he joins many others (including myself) that have taken a review approach or have collated material under the legal geography umbrella (see, for example, Blomley et al. 2001; Bartel et al. 2013; Braverman et al. 2014; Jeffrey 2019; Orzeck and Hae 2019; O'Donnell et al. 2020a). In "Legal Geography I", Delaney outlines a growing scholarship, covering vastly disparate topics, landscapes and social relationships that reveal the innate complexity in the co-constituted processes that law-place-people connections create (Delaney 2015). In "Legal Geography II" Delaney (2016) takes us to a legal geography literature that is underpinned by an often-unspoken commitment to revealing justice/injustices in place-law dynamics. Here we link to well-established notions within human geography of spatial injustice/s (see, for example, the classic work of Harvey 1973 and Soja 2010). In this paper Delaney claims that critical geography brings an analysis of power and power relations to questions of spatial justice, and that some key legal geography scholarship is imbued with this perspective (e.g. Blomley various). Simultaneously, Delaney argues that some progressive legal scholars have produced work that aligns strongly with this sentiment; and which introduces law faculty and students to a more reflective, critical positionality (e.g. Harm-Benson 2014; Pruitt 2014). Finally, in the third paper of his review, Delaney (2017) looks to the future. He identifies how both established and emergent literature embraces new themes and links with other cognate disciplinary and conceptual thinking. Delaney looks at work beyond the 'Western gaze' that characterises so much of the existing legal geography literature. He looks at the influence of international regimes and articulates legal geographical scholarship with the physical, more (or other) than human worlds. Importantly, Delaney highlights scholarship from beyond North America and the UK. Coming on the heels of these review papers, Jeffrey (2019), also publishing in *Progress in Human Geography*, takes a similar approach in his work which reviews legal geography vis-a-vis court-room materiality—the architecture of law's practice in a litigation setting—to engage in an explicit way with law's spatiality. Extending legal geography's practice to include new sites, new peoples and to redress structural problems is pushing the discipline forward in exciting ways.

Other, selected, and more recent critical works in an emergent legal geography scholarship that extends the remit of the legal geography praxis include Braverman and co-workers, *The Expanding Spaces of Law (a timely legal geography)* (2014). In this edited collection the authors bring together scholars from law and geography, again based largely in the Global North, to consider a range of issues through a legal geography lens. The spatial dimension of the legal geography project sits at the fore of this work, though the editors do note an emergent scholarship which links other key geographical components (time and power) to the project. While geographers have long paid attention to time (through historical geography in human geography, and biogeography, geomorphology and Quaternary studies in physical geography), in *legal* geography scholarship this has not necessarily dominated, or even been clearly expressed, in research agenda (following Valverde 2015). Others in the collection focus on ideas of comparative legal geography (Kedar 2014), taking legal transplants that links with ideas from legal pluralism (which I consider in paragraphs to follow). Throughout this recent work there is a clear and growing awareness of the usefulness of legal geography perspectives in translating regulatory obligations across scalar divides. The chapters in Braverman's collection draw out some interesting new areas for legal geography scholarship, including 'everyday practices' (Azuela and Meneses-Reyes 2014), habits (Blomley 2014), rural legal geographies (Pruitt 2014), methodologies (Braverman 2014) and litigation (Harm-Benson 2014). The editors (Braverman et al. 2014) do note, as do other key review papers, that

> (w)e believe that the legal geography project would be enriched by studies situated out of the usual ambit of the largely urban, Global Northwest. We also think that legal geography will prove a useful tool in marginalized contexts. (p. 9)

In part reflecting and in part responding to these calls for a more broadly constituted legal geography, research from outside the old-world bastions of the global north has emerged, and particularly so in Australia. Australian legal geography has been at the forefront of many new evolutions in the braided approach identified by Braverman et al. (2014)

to research and scholarship (see for some limited examples, Bartel et al. 2013; Delaney 2017; O'Donnell et al. 2020a). Making visible the hand of the law and making visible the role of space/place is an essential component of this work. Regional legal geography scholarship, especially emanating from the Australasian region is now pushing at the boundaries of traditional approaches. In our edited collection, myself and co-editors (O'Donnell et al. 2020a), explore a distinctive environmental legal geography explicitly through methods and methodology. Using a range of places, peoples and laws/lores, from Australia, New Zealand and the Asia-Pacific region the authors in this edited collection describe and critique the law-place-people trinity. We are enabled through this research work to learn about regulatory place-making beyond the traditional Anglo-American perspective. This builds upon work from regional based scholars, including but not limited to Sherval (2020), Graham (various), Bartel (various), Turton (2015a, b, 2017, 2020), Robinson (various), O'Donnell (various), O'Donnell et al. (2020a, b, c), Jessup (2013, 2015) and my own work Gillespie (various) which showcase their diverse research foci and collaborations across the region. What is revealed in this array of work is not a linear and causative 'law/lore + place = landscape' equation. Rather, legal geography scholarship may reverse that equation or recombine it in unexpected ways. The configurations are endless. They are circular, repetitive and infinite. What is clear is that place and law are deeply and inescapably entangled.

3.4 Place and Material Considerations

Ideas surrounding 'place' are often centre stage in legal geography scholarship and the literature is replete with case studies articulating the messiness of people-place-law interactions (for example, Bartel and Graham 2016). Place is, of course, spatially and temporally contingent. Place-based understandings of landscape, which require vernacular knowledge (Bartel 2014), often sit uneasily with perceptions of landscape held by others, elsewhere. As a consequence, regulatory burdens imposed by administrators in far-away locations may chafe against, or flatly contradict, place-based experience and understandings (Della Bosca and Gillespie 2019). This innate tension in the people-place-law connection is a fruitful arena for investigation by legal geography scholars that will underwrite improved policy solutions.

As a way of understanding how this might work in practice, we can take the intractable problem of biodiversity loss (see Chapter 2) as an example. In a standardised policy response (Fig. 3.1), the habitat of species or groups of species is protected by creating a protected area. This protected area is emplaced or overprinted upon existing land and waterscapes and restrictions on human usage/s must be imposed to halt degradation. In the monitoring stage it is found that the environmental protection regime is failing (regulatory failure) because there is non-compliance with the

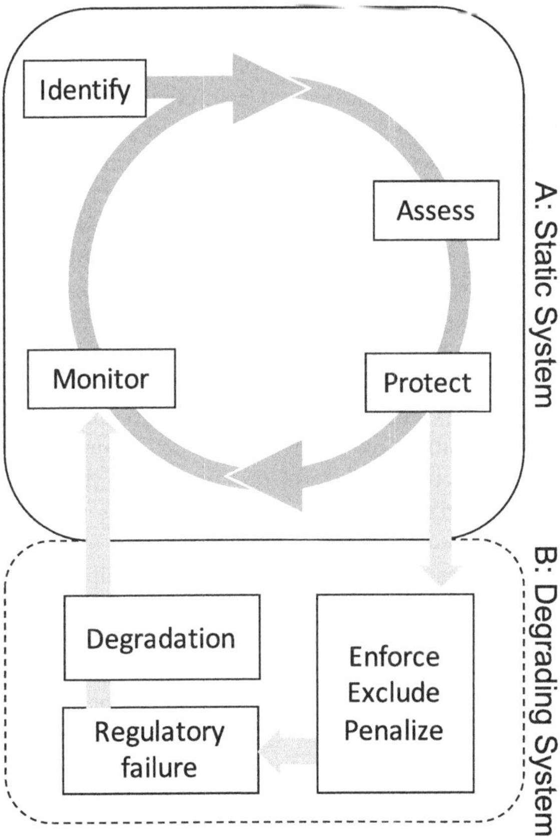

Fig. 3.1 Schematic environmental problem—identification to regulatory failure phases

laws created to manage or exclude human use. In order to avoid this commonplace scenario, I argue that if we adopt a legal geography approach at the beginning of the cycle—before the regulatory response is developed and emplaced—then the regulatory solutions can be more appropriately tailored to avoid non-compliance, regulatory failure and, in this case, further biodiversity loss.

Understanding how people live, work and go about daily tasks in particular locations helps us to understand both negative and positive place attachments (Bartel and Graham 2016). Attachments can act to alienate; attachments can also be reinforced with the power of law. Della Bosca and Gillespie (2019), for example, identify how some 'local' views can be privileged in environmental planning processes. This work demonstrates that the level or nature of participation in environmental decision making is not equal between locations or between things. Running public meetings for development applications in local venues allows local voices to be heard—generally considered a positive part of 'good' environmental decision-making and participatory planning process (in a deliberative democracy). However, this physical process excludes many other voices, from voices concerned with the emissions of greenhouse gases from fossil fuels causing climate change through to the especially fragile ecosystem habitats and species that are protected through national environmental protection legislation. Legal processes can act against local interests of all living (and non-living) entities when these interests are given voice but are overshadowed but competing, louder and more dominant, stakeholders—what Lane and Corbett (2005) term the 'tyranny of the local'. As we will see in coming chapters, this perspective is incredibly valuable in describing and evaluating protected area programs where stakeholder conflicts exist between humans and living and non-living components within the protected area.

Bartel (2017) uses a blended word—a 'plegal'—to label and reinforce the interdependent relationship between place and law. Understanding the need to hear quieter voices in environmental management debates is a theme that comes through Bartel's legal geography work. Bartel's (2014) articulates well the value of capturing what she describes as 'vernacular' knowledge to clarify how law is understood by different stakeholders—how they value state-based laws. This work, drawing on farmer interviews in New South Wales, Australia, describes how farmers feel government regulation fails because they do not take adequate account of localised conditions. She writes

> There is an epistemic distance between the bureaucratic knowledge held by government and the vernacular knowledge (place-based knowledge) of heterogeneous environments held by farmers. Incorporating vernacular knowledge so that laws are more geographically sympathetic may close vernacular disjunctures and cure regulatory failure.

My own work reinforces this observation: laws do not work (as well) as they are intended to when they do not resonate with either the people or land/waterscapes upon which they are imposed (Gillespie various). The need to capture vernacular knowledges is, I argue, essential for good regulatory drafting and implementation. Part of this call is to recognise the value in adopting an holistic approach to management issues. Linking natural scientific and social scientific knowledge to environmental management problem-solving is increasingly essential (see, for example for commentary on global trends Aswani et al. 2018 or for a regional example, Penny et al. 2016). Again, this approach has the potential to enhance protected area management techniques—as has become well recognised by many international conservation organisations (for the IUCN approach see Cross et al. 2017).

We have described legal geography as interested in the circular, spiralling connection between law and place, where each becomes braided in a complex, interlaced structure. To emphasise this, we now turn to consider the phenomenon broadly described under the 'localisation of law' umbrella within the legal geography rubric. While law imposes itself on land and water scapes and the people who inhabit them, these places also influence the way that law is performed. Place and law fold into and over one another, twist around each other to create an intertwined structure that is impossible to disentangle without careful attention to the detailed and often fragile connections holding it together. Bennett (2016) provides a useful analysis of the 'localisation of law' idea,

> the existence of 'as-made' generic law does not automatically shape the making of a place *necessarily*. Instead, there must be some process of localisation by which that provision has been brought to or asserted upon a site. Localisation will often also entail an adaptation to the specificity of local conditions – neither laws, ideas nor material forms will simply transplant 'as is' to their further sites, there will be mutations (which in the case of place, gives it its characteristic specificity, whilst still retaining some resemblance to the form from which it has derived). (p. 184)

Paying attention to place creates an obligation to consider the material. The 'material', defined narrowly in this setting as land and waterscapes (the material world), imposes itself systemically on the nature and performance of law/s and lore/s. Much work in legal geography describes the ways in which law imposes upon the material (see Nagle 2010) but, equally, it is important to recognise that the material imposes upon law, for places can shape regulatory responses. Material land and waterscapes are not necessarily isolated. Places can be rural and isolated, but they are often, sometimes only briefly, peopled. To understand the role of each component part in shaping and re-shaping places is an essential task of the legal geographer (see also Graham et al. 2017).

3.5 Property and Territory

Another critical component of legal geography scholarship has been to critique the notion of property. One Australian-based legal geographer interrogating the impacts of the property-place paradigm is Nicole Graham. Graham's work argues that it is the property regime that enables catastrophic environmental degradation, because this system of thinking and way of administration abstracts the connection between people and landscape and 'de-physicalises' places. She writes

> It is necessary, therefore, to interrogate the concept of property and its associated vocabulary of dephysicalisation and entitlement. Such an inquiry will make possible the introduction or return of the concepts of locally and physically relevant land laws within a discursive structure of responsibilities and place-based knowledge. Property regimes, or land laws, can only be viable, authoritative and enduring where sufficient knowledge and responsiveness to the capacities and limits of the specific places of jurisdiction are developed and exercised. (Graham 2012, p. 97)

The appeal of the dephysicalisation concept is strong. The idea is common-sensical, applied and intuitive for legal geography scholars. Graham's work has gained prominence and traction for this essential observation.

For legal geography the relationship between people and place finds voice through an interrogation of regulatory regimes. In doing so, the idea of territory vis-à-vis property surfaces. Some of the most influential work in this area comes from one of legal geography's scholarly leaders, Nicolas Blomley. For the best part of two decades Blomley's work has emphasised the importance of property relationships (see, for selected examples, 2004, 2005, 2016a, b, 2019). From describing bounding and excluding processes in a variety of contexts (see, for example, Blomley 2004, 2005), Blomley's emphases the need for geographers (in particular) to think and re-think about the ways in which property, far from being passive or benign, is deeply implicated in place-making through law's regulating reach. This has been echoed by others (see, for example, the work of Holmes 2014 for an Australian context). Thinking about and documenting the complex ways people connect with and often alienate landscapes (and waterscapes) to control and exclude through property (law) processes, enables us to better understand the mutual imbrication of law and space (Delaney 2015). Understanding connections to land and property (and defining property in the 'ownership' or 'Blackstonian' model (Blomley 2005) not as the object but as a social—people to people—relationship) then equally understanding the processes of bounding, buffering, possessing, excluding and asserting use rights becomes essential in implementing protected area programs. We return to these concerns in the following chapters.

Discussion of property rights is intimately linked with the need to better understand territory (Delaney 2005). In bringing territory into focus, we begin to see clearly how power dynamics are intimately linked to territorial reconfigurations. Blomley's insights into the idea of territory draws on the work of geographer Stuart Elden in this respect (for example see, Elden 2005, 2010, 2013a, b). Elden argues for a focus on the idea of territory and draws our attention to the need to think beyond 'land' or 'terrain' in order for the idea of territory to be understood in its historical and place-based (locational) settings. The political elements of this work are important and clearly influence the related legal geography literature (see, especially Blomley 2019). This work draws our attention to issues of power; how power is created and held through time and, critically, is aided through legal processes. Blomley (2019) takes us to seventeenth century rural England to exemplify practices of exclusion which resonate to this day. In a similar vein, these sentiments are echoed in drawing our attention to power dynamics through regulatory processes associated with territorialisation and property in Graham's close and detailed examination of property law's evolution in an Australian context. In *Lawscape*, Graham (2010) reaffirms the law's position: property is place-less. It is not an idea connected to land or waterscapes but it is formalised through law to control arrangements between people, which disconnects us from places. These central observations enable us to untangle law's complicity in crafting rules that are disassociated with place-based concerns and are important considerations to return to in our assessment of protected area management approaches. We will return to these concerns in later chapters.

3.6 Conceptual Nuance: The Call for a Justice Inflected 'Legal Ecology'

Legal geography scholarship has been criticised strongly for an apparent lack of engagement with a theory or concept; the focus, it was argued, was too applied, too pragmatic, too case-study based and lacked any overarching explanatory framework. Whatever the merit of these criticisms, the situation has certainly changed in recent years. In the following section I will take up two conceptual underpinnings that highlight how power dynamics can be linked to legal geography work beyond property and territory frames. In this section I want to highlight the efforts of those generating and calling for new energy in these areas. This call is also incredibly useful in our attempts to reveal the 'other' in newer forms of legal geography research.

Much, if not all, legal geography research since its emergence in the 1980s demonstrates an underlying commitment to ideas around justice, fairness and transparency—concepts also deeply embedded in legal and socio-legal scholarship. As another early stalwart in the legal geography trope, Clark, observed in *The Legal Geographies Reader*, "(T)heoretical virtue is no substitute for political and institutional commitment to the welfare of others" (2001, p. 114). One key element in some legal geography research is a determination to unravel regulatory dynamics to reveal complex people-place *power* interactions. It is for this reason that some commentators (Andrews and McCarthy 2014; Salgo and Gillespie 2018; O'Donnell 2019) point to the need for a greater consideration of a long-established geographical approach—political ecology—to complement our work.

Tracing how, where and by whom decision making crafts land and waterscapes represents an exercise in doing political ecology. The idea of a 'political ecology' emerged from the prominent work of Piers Blaikie, *The Political Economy of Soil Erosion in Developing Countries* (1985). While the field is vast, Bryant and Goodman (2008) point to some key elements that characterise the approach that emerged from Blaikie's work, three of which are relevant here: (1) the need to recognise the work of scientific experts in the natural sciences and then to extend this to consider social and political conditions; (2) the need for post-structural analysis wherein a variety of perspectives are used for causal explanations, and (3) to apply these idea/s to a wide array of topics and settings, beyond development or area studies. To this end, in my own research work I have used the framing of political ecology to explain power dynamics in decision-making processes for bushfire management in the Blue Mountains region of New South Wales, Australia (Salgo and Gillespie 2018). In this work we trace the changing role of the State Government's regulatory regime for vegetation clearing in bushfire-prone areas by, inter alia, examining parliamentary debates which were clearly influenced by a valorisation of private property entitlements. Other legal geographers have taken a similar path. O'Donnell (2019) makes a case for adapting the broad political ecology approach to enable legal geography work to consider the political-legal nexus in landscape management. Using rapidly changing NSW coastal planning processes as the example, O'Donnell highlights how coastal private property (landowner) interests can influence coastal management regulations and demonstrates in this work how a detailed account of a rapidly changing legal landscape is intimately connected to underlying political manoeuvring. Both Salgo and Gillespie (2018) and O'Donnell (2019) reference the call by Andrews and McCarthy (2014) to explicitly link legal geographical research to political ecology.

Legal geography research inflected with political ecology focusses our attention on power relationships in people-place-law encounters. In an environmental governance context, decision-making processes—defined through legislative and case law obligations (i.e. the law)—reveal how powerful interests (e.g. stakeholders, interest groups) can use regulations to change the ways we interact with land and water scapes. One clear way that this analytical lens can sharpen our focus occurs when these perspectives reveal inequalities; the issue of gender and gendered disadvantage is one example. I argue that a legal geography lens, imbricated with political ecology, can work to unpack gendered disadvantage.

Feminist legal geography is a rapidly growing area of legal geographical work (Brickell and Cuomo 2019; Cuomo and Brickell 2019). Adopting the labelling "geolegality" (from Smith, 2014), Brickell and Cuomo set out a compelling case for thinking about the spatial-scalar geo-legal through a gendered lens in "a project that integrates the intellectual terrains of legal geography and feminist geopolitics" (p. 105). Their work sets out to demonstrate that a feminist geolegality framing allows us to see how gender power dynamics range across scales—from the intimate to global. They argue that this is a necessary step in creating a more thorough geolegal intellectual endeavour. One compelling point made by Brickell and Cuomo, which resonates with work aiming to show how disadvantage and discrimination can cut across identities (including the disrupting impacts of protected areas, see Chapter 2), is that we can take the gender/power/law nexus further, toward an intersectional (identity based)/power/law nexus (p. 107). In an effort to push us toward re-orienting our thinking about power, decision-making and multiple scales and identities, the feminist geolegality perspective provides us with a valuable tool. I have embraced this attitude and positioning in my own work (Gillespie and Perry 2019; Perry and Gillespie 2019). We report our research findings in these papers which document the law/power/gendered dynamics of protected wetlands management. We return to consider these concerns in Chapter 5.

Being alert to the ways in which social injustices are manifest in regulatory processes can reveal how legal geography scholarship is underpinned by a search for justice. As O'Donnell points out, "… legal geographers aim to pursue lines of legal-spatial enquiry in a quest for social and environmental justice" (O'Donnell 2019, p. 128). Attempting to rectify incidents of social vulnerability are a key component of political ecology (Forsyth 2008) and the need to acknowledge the role of different perspectives—from scientific to social to more-than-human value/s—to inform a more just and equitable environmental policy response underscores this approach.

3.6.1 Not Forgetting the Value of Environmental Justice

A cognate endeavour is also taking place in our legal geography scholarship that parallels the evolution and cross-connection between legal geography and political ecology, and which is also fundamental to the approach adopted in this book. As I've just written, understanding power dynamics using a political ecological lens is pivotal in unravelling complex socio-ecological connections, but equally important is the need to account for key concepts that find their foundation in law (at least superficially)—how justice, fairness and transparency exist, or fail to—in the law-environment encounter. To this end, some legal geography scholarship explicitly calls upon environmental justice frames. Delaney (2015) picks up this theme in his review paper (described earlier) wherein we note his observation:

> (e)very instance of injustice implicates power, and any occasion of power can be evaluated in terms of justice. But they are distinguishable. Power talk is descriptive, justice talk is normative. To identify a state of affairs as unjust is to condemn it. In our world, much of what we recognize and experience as power is directly linked to the legal – to rules, rights, authority, punishment and so on. (p. 268)

Environmental justice begins to assert itself. Environmental justice is both a social movement and a concept. With roots in racial/spatial injustices associated with vulnerable communities and social movement activism in the United States (Walker 2012), environmental justice is also a very useful conceptual device. Environmental justice is generally defined as seeking justice through decision-making that has four inter-related core components: (1) Distribution, (2) Recognition, (3) Capabilities (drawing on Sen 2005 and Nussbuam 2003) and (4) Participation (Schlosberg 2007). In calling for environmental justice, or in seeking it through research programs and scholarly activities, we are seeking to investigate who gets what, who misses out and who is adversely or unfairly forced to cope with degrading environmental conditions (through polluted water sources, by way of one example). But who exactly is the 'who' in these questions? Those people adversely impacted by power imbalances in critical decision-making roles—drawing on observations from political ecology? Without doubt these peopled communities are implicated. But so too are the more-than-human communities who are often quieter voices in debates. However, for the moment we consider the changing policy objectives that seek to create more just conservation decisions.

There is a clear and distinct focus on social (human) environmental in/justice/s in conservation policy circles. Martin et al. (2016), explicitly reinforcing Aichi Target (Strategic Goal C, Target 11) to manage protected areas equitably, look to concerns about a lack of social justice for local communities in conservation practice. They take up recognition as a necessary component part of equity in their call that

> … conservation should look beyond a distributive model of justice to incorporate concerns for social recognition, including careful attention to ways to pursue equality of status for local conservation stakeholders. This will require reflection on working practices and looking at forms of intercultural engagement that, for example, respect alternative ways of relating to nature and biodiversity. (Martin et al. 2016, p. 254)

Much work in protected area management has focussed on the need to take better account of local voices, interests and futures (see also the recent work of the IUCN's engagement with indigenous voices in Cross et al. 2017). We turn to consider these issues further in Chapters 4 and 5 with a focus on world heritage and wetland protected areas.

Shoreman-Ouimet and Kopnina (2015) call for conservation practice to adopt a middle ground between advancing human and non-human interests. It is a call to social scientists working in conservation settings to shift from an anthropocentric to ecocentric perspective—taking their passion for justice beyond humans. Their call is about better recognition of non-humans. In an insightful paper Washington et al. (2018), develop a position that calls for the academy (especially conservation-interested scholars) to foreground their work in ecological justice or "eco-justice". Eco-justice builds on social and environmental justice and thus aims to not exclude justice-for-humans as they argue some ecocentric perspectives might suggest. This position has important implications for the approach we take in protected area management vis-à-vis restricting human activities.

3.6.2 And Ecological (In)Justice

Taking one element of environmental justice—recognition—we turn to focus on justice outside the tradition social/human frame and consider this in the context of justice for the more-than-human, often considered to be aligned with *ecological* justice (Schlosberg 2007). Baxter (2004) articulates this concern well, arguing for the need to consider a variety of organisms in justice claims as our current approaches—moral and regulatory—have proven unsatisfactory. Baxter argues for furthering an ecological justice agenda wherein life-forms other than humans can make a (distributive) justice-based claim which recognises they are entitled to a "fair share of the environmental resources which all life-forms need to survive and to flourish" (Baxter 2004, p. 4). Baxter recognises that through institutional and governance efforts humans across the globe have taken serious steps to maintain ecosystems and advance justice claims, and points to the five international environmental instruments (including the *World Heritage Convention* and the *Ramsar Convention*) and national commitments to these instruments as some evidence of the commitment. He also points out that we need to go further to give fuller recognition (e.g. Constitutional, or national legislative protection) for non-human species.

In an environmental protection and decision-making context, the role of law to regulate human activities so as to not undermine species and habitat conservation (that is, to give non-human species recognition) is critical to the success of the protected area program across the globe. Without law/s to identify threatened, vulnerable and endangered species and habitats and to control human (ab)uses, degradation will accelerate. In an Australian context the prominent national environmental law, the *Environmental Protection and Biodiversity Conservation Act 1999* (Commonwealth), clearly aims to stem biodiversity losses (s.3, s.s.1(c)). However, the actual success of the law in preserving and conserving biodiversity is debatable. Jessup (2015), using a legal geography lens and principles adapted from environmental/ecological justice, demonstrates the way in which environmental law does not work.

Investigating a dispute over a proposal to log Tasmania's Wielangta Forest, Jessup takes up the justice-as-recognition cudgels for non-human interests (Jessup 2015). He argues ecological injustices "... occur when participants, subjects or features of environmental disputes are ignored, are overlooked, or their interests are downplayed" (p. 5). Jessup analyses a case (*Brown v Forestry Tasmania [No 4]*), involving three protected species under the national legislation, the *Environmental Protection and Biodiversity Conservation Act 1999* (C'th), being the Tasmanian wedge-tailed eagle, the broad-toothed stag beetle and the Swift parrot. Despite clear environmental law for protection of these species, the lack of voice given to the more-than-human disturbed Jessup

> I could not find the justice in this case. However, it was not access or participation, or indeed distribution, which stood out as a gap in the laws in this instance, rather there was another unfairness evident in the case: unfairness at not even contemplating the interests of the non-human species at the centre of the dispute. (p. 8)

Speaking for the more-than-human is essential in the context of protected area management which aims to stem devasting biodiversity losses. As Jessup suggests, law's remit to protect must extend to giving voice to those it seeks to protect.

Another working within a legal geography tradition who collapses the nature/human binary in her work, and which attends directly to conservation and preservation issues, is Irus Braverman (see, for example, 2012, et al. 2014, 2015a, b). In *Wild Life* (2015a), Braverman considers the problem of a protected area approach through the lens of the in situ and ex situ question: should species be preserved in captivity or should they be saved in the 'natural' habitat? In interviews with a range of professionals both for and against ex situ conservation practices, Braverman narrates an intricate co-dependency between two world views about the best way to preserve a range of non-human species in an era of rapid biodiversity decline. A key question revolves around how humans will manage species in perpetuity; whether there exists the capacity over time to aim for 'natural' restoration of species in situ for functional ecosystem equilibrium with (minimal) human management. The focus is on newer ways of thinking about the complexity of managing biodiversity in an age of change. Braverman's approach is not exclusively regulatory, though she notes that "(l)aw, in its myriad forms and variations, plays a central role … a more nuanced account of the variety of legalities at work in conservation" (p. 229). This work draws our attention to the ways in which legal strategies both allow the in situ/ex situ divide to continue but also enable it to mutate. Critically, Braverman (2015a) shows us that a legal geography approach, in making visible the hand of law, reveals complexity that is not at first glance easily discernible. The book unpacks "… various ways in which the legal protections assigned to individuals, populations, and species are applied and constantly negotiated between in situ and ex situ" (p. 230). In *Animals, Biopolitics, Law: lively legalities* (2015b), Braverman further delves into the regulatory settings for more-than-humans in describing the phenomenon of protected species/habitats lists. These lists act to classify and order species—some are vulnerable, while others are threatened or endangered. The IUCN Red List for Threatened Species is famous worldwide and includes some of the world's most iconic and recognisable flora and fauna (IUCN Red List, undated). This preferencing of iconic species for protection highlights law's role in regulating via hierarchies, of playing favourites and of elevating some species so as to give them voice. Law's systemic ordering through lists preferences the sorting, classification, ranking, assessment and monitoring innate to conservationists (from the professional ranks of e.g. biology, ecology etc.). The listing sorts things we value to greater and lesser degrees. This process formalises favour and privileges. This does not promote a relational, hierarchy-less understanding of place-people dynamics.

Recognition for multispecies—indeed creating multispecies justice—encourages new ways of thinking about and seeing the world around us. It is a theme of geographical scholarship through the ages, with research from eminent scholars such as Haraway (e.g. 2018 but note that there are many, many other examples from a vast and growing literature) leading the way. Moving towards a form of spatial justice with responsibility for 'others' is evident in much relational geographical scholarship. Bawaka Country (Country et al. 2016) demonstrate how we can give voice to country. In a novel approach that exposes the limitations of western thinking this work demands us to re-think the place-people nexus as being both co-dependent and emergent, as the author's explain: "... we look to Country for what it can teach us about how all views of space are situated, and for the insights it offers about co-becoming in a relational world" (p. 455). They explain the use of the term country to include "... humans, more-than-humans and all that is tangible and non-tangible and which become together in an active, sentient, mutually caring and multidirectional manner in, with and as place/space" (p. 456). Inhabited land and waterscapes have agency, knowledge and power. This links explicitly to ideas emergent in ideas surrounding the rights of nature often referred to as Earth Jurisprudence.

One key proponent of the Earth Jurisprudence approach is Peter Burdon (see, for examples, 2010, 2012, 2014). This view of intrinsic rights for nature documents the degrading anthropocentric position of Western environmental law, building on the early work of Sax (1970) and Stone (1972) and questions the dominant anthropocentric approach, in favour of an ecocentric mandate. This work has received endorsement in recognition of the intrinsic rights of nature emerging throughout the world—from South America (Bolivia and Ecuador) to New Zealand (Chapron et al. 2019). A critical legal geography approach could embrace the evolution of this conceptual and applied environmental rights approach (and linking to thinking more clearly about the links between environmental protection and human rights has tentatively begun in my own work, see, for example, Gillespie 2013, 2017a, b). In addition, calls for a critical physical geography scholarship (Lave et al. 2014; Tadaki et al. 2015) to re-orientate the role of place dynamics in the Anthropocene are emerging (Delaney 2017; Knitter et al. 2019). Adding *legal* to this call for a genuine critical engagement to become a ***critical physical legal geography*** is imminent.

3.7 Outside the Dominant Paradigm

Efforts to refocus our gaze into different places and onto different things now permeates legal geography scholarship. Using a non-human lens, we are able to consider the law/place nexus for places beyond people to consider more-than-human species and habitats that shape our world. The need to critically reflect upon different ways of thinking and conceptualising what law, and lore, is, has become paramount in an era of grave biodiversity loss and ecological crises. Our locational gaze now takes us outside the dominant Anglo-American focus of much previous legal geography research. In this section we will consider how legal geography scholarship has a strong association with legally plural approaches.

Much of my work tackles the dominant legal geography narrative embedded in the Anglo-American system tradition through emergent research emanating from the Australasian region. Writing about the jarring impact of applying a classical property model based on Anglo-American property law notions—with rights to possess, bound and exclude property—in a Southeast Asian setting I have found that exportation of these ideas without critique produces uneven (and undesirable) impacts (Gillespie 2016). Not taking into account localised social connections and expectations, often embedded in lores (customs and traditions) and not laws, creates an incongruent property narrative in different places. This work decentres the dominant property narrative as it does not fit well within non-Western visions of world-making (see also Graham 2010). Beyond the disjuncture caused with inappropriate property regulatory regimes, I have also worked in other contexts in Southeast Asia in and around villages within and adjacent to protected wetland areas (Gillespie 2012, 2016, 2018, 2019; Gillespie and Perry 2019). We return to explore these concerns in later chapters.

Efforts to re-focus within legal geography scholarship have blossomed in recent years. Robinson's work, in tandem with colleagues, in intellectual property through Southeast Asia (2014, 2017), the Pacific (Robinson and Forysth 2016) and Australia (Robinson and Raven 2017, 2019) provides examples. Robinson's investigation of bio-piracy, intellectual property rights and local ecological knowledge/s draws on a locational context away from Europe and North America. Much of this work attends to connections with legal plural scholarship in an engaged and explicit manner (Robinson and Graham 2018) as it recognises the significance of regulatory worldviews from perspectives *other than* the Westphalian liberal democratic states in the Anglo-American tradition. The entanglement between legal geography and legal pluralism influences how we assess the success of protected area regimes (Gillespie 2012). Another prominent legal geographer, Sandy Kedar (with colleagues) also looks to other places to reveal law's complicity in shaping landscapes beyond the traditional Anglo-American remit (see especially Kedar et al. 2018 and see also Calyx et al. 2020). Beyond the western gaze, these authors encourage us to refocus on the complexity of historical, legal and territorial disputes in the southern Israel/Palestinian region. They reveal the layering of laws in place where indigenous property and settlements systems continue to exist with/in state systems.

3.8 Conclusion

Environmental protection regimes are crafted through law and, perhaps to a lesser extent, lore. These rules, whether state or informal in character, are context and place specific. It is for this reason that a legal geography approach enables greater clarity in identifying, describing and explaining the success or failure of our current governance arrangements. A strong scholarship has emerged in this field that throws light upon structural concerns, especially the influence of property and territory, in shaping the places we live and work.

Separating, literally and conceptually, people from their physical environment through the prism of private property control is inherently problematic. The insistence on rolling-out private property regimes born from liberal Western models onto different places is profoundly problematic. The decolonialisation agenda requires that we re-think the value of the roll-out. Turning a spotlight on to the imperfections, power imbalances and injustices within the regulatory regime of property/environment/place interactions continues to be a central concern within legal geography research. Giving greater voice to a variety of stakeholder perspectives, from indigenous and local communities and to the more-than-human, demands this continual critique.

Critically, legal geography scholarship does not sit in isolation from several important scholarly traditions. This chapter demonstrates the value in blending or combining legal geography with other thematic and analytical areas in geography and law, particularly political ecology and environmental justice. Linking to political ecology and environmental justice lends our descriptive analysis of law's impact and vice versa (society's impact on law) a clearly critical lens. This creates a hybridised legal geography—I have coined the term legal ecology—that draws together and extends concerns with power and justice. In this, legal geography scholarship becomes a conceptual and analytical platform empowering its adopters to tackle with confidence the complexities of environmental regulation in the Anthropocene.

In the next chapter we consider one type of protected area that exists by virtue of one of our key international environmental instruments, the *World Heritage Convention*. An application of the key precepts from legal geography enables us to consider how World Heritage protection regimes work. We see regulatory layers in formal protection obligations from the global to the local and from the Western legal gaze through non-Western local lore and practices. Our perspective on the success of this lauded conservation initiative changes through the prism of legal geography, for we gain clarity about the benefits and shortcomings of the existing conservation regimes.

References

Andrews E and McCarthy J (2014) Scale, shale, and the state: Political ecologies and legal geographies of shale gas development in Pennsylvania. *Journal of Environmental Studies and Sciences* 4(1): 7–16.

Aswani S, Lemahieu A, Sauer WHH (2018) Global trends of local ecological knowledge and future implications. *PLoS One* 13(4): p. e0195440.

Azuele A and Meneses-Reyes R (2014) The everyday formation of the urban space: Law and poverty in Mexico City. In: Braverman I, Blomley N, Delaney D and Kedar A (eds) *The Expanding Spaces of Law: A Timely Legal Geography*. Stanford University Press, Stanford, pp. 167–189.

Bartel R (2014) Vernacular knowledge and environmental law: Cause and cure for regulatory failure. *Local Environment: The International Journal of Justice and Sustainability* 19(8): 891–914.

Bartel R (2017) Place-thinking: The hidden geography of environmental law. In: Philippopoulos-Mihalopoulos A and Brooks V (eds) *Handbook of Research Methods in Environmental Law*. Edward Elgar Publishing, Cheltenham, UK.

Bartel R and Graham N (2016) Property and place attachment: A legal geographical analysis of biodiversity law reform in New South Wales. *Geographical Research* 54(3): 267–284.

Bartel R, Graham N, Jackson S, Prior JH, Robinson DF, Sherval M and Williams S (2013) Legal geography: An Australian perspective. *Geographical Research* 51(4): 339–353.

Baxter B (2004) *A Theory of Ecological Justice*. Routledge Research in Environmental Politics, Routledge, New York.

Bennett L (2016) How does law make place? Localisation, translocalisation and thing-law at the world's first factory. *Geoforum* 74: 182–191.

Blaikie P (1985) *The Political Economy of Soil Erosion in Developing Countries*. Routledge, Oxford and New York.

Blomley N (2004) The boundaries of property: Lessons from Beatrix Potter. *The Canadian Geographer* 48(2): 91–100.

Blomley N (2005) Remember property? *Progress in Human Geography* 29(2): 125–127.

Blomley N (2014) Learning from Larry: Pragmatism and the habits of legal space. In: Braverman I, Blomley N, Delaney D and Kedar A (eds) *The Expanding Spaces of Law: A Timely Legal Geography*. Stanford University Press, Stanford, CA, pp. 77–94.

Blomley N (2016a) The territory of property. *Progress in Human Geography* 40(5): 593–609.

Blomley N (2016b) The boundaries of property: Complexity, relationality, and spatiality. *Law & Society Review* 50(1): 224–255.

Blomley N (2019) The territorialization of property in land: Space, power and practice. *Territory, Politics, Governance* 7(2): 233–249.

Blomley N, Delaney D and Ford R (2001) *The Legal Geographies Reader*. Blackwell, Oxford.

Braverman I (2012) *Zooland: The Institution of Captivity*. Stanford University Press, Stanford, CA.

Braverman I (2014) Whose afraid of methodology? In: Braverman I, Blomley N, Delaney D and Kedar A (eds) *The Expanding Spaces of Law: A Timely Legal Geography*. Stanford University Press, Stanford, CA, pp. 120–141.

Braverman I (2015a) *Wild Life: The Institution of Nature*. Stanford University Press, Stanford, CA.

Braverman I (2015b) *Animals, Biopolitics, Law: Lively Legalities*. Routledge, Oxford and New York.

Braverman I, Blomley N, Delaney D and Kedar A (2014) *The Expanding Spaces of Law: A Timely Legal Geography*. Stanford University Press, Stanford, CA.

Brickell K and Cuomo D (2019) Feminist geolegality. *Progress in Human Geography* 43(1): 104–122.

Bryant R and Goodman MK (2008) A pioneering reputation: Assessing Piers Blaikie's contributions to political ecology. *Geoforum* 39(2): 708–715.

Burdon P (2010) Wild Law: The philosophy of earth jurisprudence. *Alternative Law Journal* 35(2): 62–65.

Burdon P (2012) A theory of earth jurisprudence. *Australian Journal of Legal Philosophy* 37: 28–60.

Burdon P (2014) Idealism and struggle: Co-opting legal rights for environmental protection. *Alternative Law Journal* 39(3): 176–178.

Calyx C, Jessup B and Sihombing M (2020) Asserting land rights through technology and democratic expression: The effect of the Indigenous Peoples Alliance of the Archipelago v Indonesia case. In: O'Donnell T, Robinson DF and Gillespie J (eds) *Legal Geography: Perspectives and Methods*. Routledge, London and New York.

Chapron G, Epstein Y and López-Bao JV (2019) A rights revolution for nature. *Science* 363(6434): 1392–1393.

Clark G (2001) Foreword. In: Blomley N, Delaney D and Ford RT (eds) *Legal Geographies Reader: Law, Space and Power*. Blackwell, Oxford.

Country B, Wright S, Suchet-Pearson S, Lloyd K, Burarrwanga L, Ganambarr R, Ganambarr-Stubbs M, Ganambarr B, Maymuru D and Sweeney J (2016) Co-becoming Bawaka: Towards a relational understanding of place/space. *Progress in Human Geography* 40(4): 455–475.

Cross R, Doornbos S, Cooney R, Wong P, Mead A, Lindeman K, Kanagavel A, Parvathy S, Tomasini S, Montanari B, Gabrys K and Kehaulani Watson-Sproa T (2017) Guidance for integrating indigenous and local knowledge (ILK) in IUCN Red List assessments. Gland, Switzerland.

Cuomo D and Brickell K (2019) Feminist legal geographies. *Environment and Planning A: Economy and Space* 51(5): 1043–1049.

Delaney D (2005) *Territory: A Short Introduction*. Wiley Blackwell, Hoboken.

Delaney D (2015) Legal geography I: Constitutivities, complexities, and contingencies. *Progress in Human Geography* 39(1): 96–102.

Delaney D (2016) Legal geography II: Discerning injustice. *Progress in Human Geography* 40(2): 267–274.

Delaney D (2017) Legal geography III: New worlds, new convergences. *Progress in Human Geography* 41(5): 667–675.

Della Bosca H and Gillespie J (2019) The construction of 'local' interest in New South Wales environmental planning processes. *Australian Geographer* 50(1): 49–68.

Elden S (2005) Missing the point: Globalization, deterritorialization and the space of the world. *Transactions of the Institute of British Geographers* 30(1): 8–19.

Elden S (2010) Land, terrain, territory. *Progress in Human Geography* 34(6): 799–817.
Elden S (2013a) *The Birth of Territory*. University of Chicago Press, Chicago, IL.
Elden S (2013b) How should we do the history of territory? *Territory, Politics, Governance* 1(1): 5–20.
Forsyth T (2008) Political ecology and the epistemology of social justice. *Geoforum* 39(2): 756–764.
Gillespie J (2012) Legal pluralism and world heritage management at Angkor, Cambodia. *Asia Pacific Journal of Environmental Law* 14(1–2): 1–19.
Gillespie J (2013) World heritage protection and the human right to development: Reconciling competing or complementary narratives using a Human Rights-Based Approach (HRBA)? *Sustainability* 5(7): 3159–3171.
Gillespie J (2016) A legal geography of property, tenure, exclusion, and rights in Cambodia: Exposing an incongruous property narrative for non-Western settings. *Geographical Research* 53(4): 256–266.
Gillespie J (2017a) Engaged enquiry in environmental law: Understanding people/place connections through a geographically informed human rights lens. In: Philippopoulos-Mihalopoulos A and Brooks V (eds) *Research Methods in Environmental Law: A Handbook*. Edward Elgar Publishing, Cheltenham, pp. 265–285.
Gillespie J (2017b) Heritage and human rights: Reframing the conservation ethic. In: Durbach A and Lixinski L (eds) *Heritage, Culture and Rights: Challenging Legal Discourses*. Hart Publishing, Oxford, pp. 165–180.
Gillespie J (2018) Wetland conservation and legal layering: Managing Cambodia's great lake. *Geographical Journal* 184(1): 31–40.
Gillespie J (2020) Challenges in legal geography research methodologies in cross-cultural settings. In: O'Donnell T, Robinson DF and Gillespie J (eds) *Legal Geography: Perspectives and Methods*. Routledge (Taylor & Francis), New York.
Gillespie J and Perry N (2019) Feminist political ecology and legal geography: A case study of the Tonle Sap protected wetlands of Cambodia. *Environment and Planning A: Economy and Space* 51(5): 1089–1105.
Graham N (2010) *Lawscape: Property, Environment, Law*. Routledge, London.
Graham N (2012) Dephysicalisation and entitlement: Legal and cultural discourses of place as property. In: Jessup B and Rubenstein K (eds) *Environmental Discourses in Public and International Law*. Cambridge University Press, New York, pp. 96–120.
Graham N, Davies M and Godden L (2017) Broadening law's context: Materiality in socio-legal research. *Griffith Law Review* 26(4): 480–510.
Haraway D (2018) Staying with the trouble for multispecies environmental justice. *Dialogues in Human Geography* 8(1): 102–105.

Harm-Benson M (2014) Rules of engagement: The spatiality of judicial review. In: Braverman I, Blomley N, Delaney D and Kedar A (eds) *The Expanding Spaces of Law: A Timely Legal Geography*. Stanford University Press, Stanford, CA, pp. 215–238.

Harvey D (1973) *Social Justice and the City*. Edward Arnold, London.

Holmes J (2014) Explorations in Australian legal geography: The evolution of lease tenures as policy instruments. *Geographical Research* 52: 411–429.

IUCN Red List of Threatened Species (undated). Available https://www.iucnredlist.org. Accessed 10 September 2019.

Jeffrey A (2019) Legal geography 1: Court materiality. *Progress in Human Geography* 43(3): 565–573.

Jessup B (2013) Environmental justice as spatial and scalar justice: A regional waste facility or a local rubbish dump out of place? *McGill International Journal of Sustainable Development Law and Policy* 9: 71–107.

Jessup B (2015) Justice, recognition and environmental law; The Weilangta Forest Conflict Tasmania, Australia. *University of Tasmania Law Review* 34: 5.

Kedar A (2014) Expanding legal geographies: A call for a critical comparative approach. In: Braverman I, Blomley N, Delaney D and Kedar A (eds) *The Expanding Spaces of Law: A Timely Legal Geography*. Stanford University Press, Stanford, CA, pp. 95–119.

Kedar A, Amara A and Yiftachel O (2018) *Emptied Lands, A Legal Geography of Bedouin Rights in the Negev*. Stanford University Press, Stanford, CA.

Knitter D, Augustin K, Biniyaz E, Hamer W, Kuhwald M, Schwanebeck M and Duttmann R (2019) Geography and the Anthropocene: Critical approaches needed. *Progress in Physical Geography: Earth and Environment* 43(3): 451–461.

Lane MB and Corbett T (2005) The tyranny of localism: Indigenous participation in community-based environmental management. *Journal of Environmental Policy and Planning* 7(2): 141–159.

Lave R, Wilson MW, Barron ES, Biermann C, Carey MA, Duvall CS, Johnson L, Lane KM, McClintock N, Munroe D, Pain R, Proctor J, Rhoads BL, Robertson MM, Rossi J, Sayre NF, Simon G, Tadaki M and Van Dyke C (2014) Intervention: Critical physical geography. *The Canadian Geographer/Le Géographe canadien* 58(1): 1–10.

Martin A, Coolsaet B, Corbera E, Dawson NM, Fraser JA, Lehmann I and Rodriguez I (2016) Justice and conservation: The need to incorporate recognition. *Biological Conservation* 197: 254–261.

Nagle JC (2010) *Law's Environment: How the Law Shapes the Places We Live*. Yale University Press, New Haven and London.

Nussbaum M (2003) Capabilities as fundamental entitlements: Sen and social justice. *Feminist Economics* 9(2–3): 33–59.

O'Donnell T (2019) Coastal management and the political-legal geographies of climate change adaptation in Australia. *Ocean and Coastal Management* 175: 127–135.

O'Donnell T, Robinson DF and Gillespie J (eds) (2020a) *Legal Geography: Perspectives and Methods*. Routledge, London and New York.

O'Donnell T, Robinson DF and Gillespie J (2020b) An Australasian and Asia-Pacific approach to legal geography. In: O'Donnell T, Robinson DF and Gillespie J (eds) *Legal Geography: Perspectives and Methods*. Routledge, London and New York.

O'Donnell T, Robinson DF and Gillespie J (2020c) Legal geography futures. In: O'Donnell T, Robinson DF and Gillespie J (eds) *Legal Geography: Perspectives and Methods*. Routledge, London and New York.

Orzeck R and Hae L (2019) Restructuring legal geography. *Progress in Human Geography*. Available https://doi.org/10.1177/0309132519848637.

Penny D, Williams G, Gillespie J and Khem R (2016) 'Here be dragons': Integrating scientific data and place-based observation for environmental management. *Applied geography* 73: 38–46.

Perry N and Gillespie J (2019) Restricting spatial lives? The gendered implications of conservation in Cambodia's protected wetlands. *Environment and Planning E: Nature and Space* 2(1): 73–88.

Pruitt L (2014) The rural lawscape: Space tames law tames space. In: Braverman I, Blomley N, Delaney D and Kedar A (eds) *The Expanding Spaces of Law: A Timely Legal Geography*. Stanford University Press, Stanford, CA, pp. 190–214.

Robinson DF (2014) Legal geographies of intellectual property, 'traditional' knowledge and biodiversity: Experiencing conventions, laws, customary law, and Karma in Thailand. *Geographical Research* 51: 375–386.

Robinson DF (2017) Traditional knowledge and biocultural diversity. In: Morand S, Lajaunie C and Satrawaha R (eds) *Biodiversity Conservation in Southeast Asia: Challenges in a Changing Environment*. Routledge, London, pp. 33–46.

Robinson DF and Forsyth M (2016) People, plants, place, and rules: The Nagoya Protocol in Pacific island countries. *Geographical Research* 54: 324–335.

Robinson DF and Graham N (2018) Legal pluralisms, justice and spatial conflicts: New directions in legal geography. *Geographical Journal* 184(1): 3–7.

Robinson DF and Raven M (2017) Identifying and preventing biopiracy in Australia: Patent landscapes and legal geographies for plants with Indigenous Australian uses'. *Australian Geographer* 48: 311–331.

Robinson D and Raven M (2019) Recognising Indigenous customary law of totemic plant species: Challenges and pathways. *Geographical Journal*. Available http://dx.doi.org/10.1111/geoj.12320.

Salgo M and Gillespie J (2018) Cracking the code: A legal geography and political ecological perspective on vegetation clearing regulations. *Australian Geographer* 49(4): 483–496.

Sax J (1970) The public trust doctrine in natural resource law: Effective judicial intervention. *Michigan Law Review* 68(3): 471–566.

Scholsberg D (2007) *Defining Environmental Justice. Theories, Movements and Nature*. Oxford University Press, New York.

Sen A (2005) Human rights and capabilities. *Journal of Human Development* 6(2): 151–166.

Sherval M (2020) Energising the law: Greening of fossil fuels and the rise of gendered political subjects. In: O'Donnell T, Robinson DF and Gillespie J (eds) *Legal Geography: Perspectives and Methods*. Routledge, London and New York.

Shoreman-Ouimet E and Kopnina H (2015) Reconciling ecological and social justice to promote biodiversity conservation. *Biological Conservation* 184: 320–326.

Smith M (2014) States that come and go: Mapping the geo-legality of the Afghanistan intervention. In: Braverman I, Blomley N, Delaney D and Kedar A (eds) *The Expanding Spaces of Law: A Timely Legal Geography*. Stanford University Press, Stanford, CA, pp. 142–166.

Soja EW (2010) *Seeking Spatial Justice*. University of Minnesota Press, Minneapolis.

Stone CD (1972) Should trees have standing? Toward legal rights for natural objects. *Southern Californian Law Review* 45: 450.

Tadaki M, Brierley G, Dickson M, Le Heron R and Salmond J (2015) Cultivating critical practices in physical geography. *The Geographical Journal* 181(2): 160–171.

Turton DJ (2015a) Unconventional gas in Australia: Towards a legal geography. *Geographical Research* 53(1): 53–67.

Turton DJ (2015b) Lawyers in Australia's coal seam gas debate: A study of participation in recorded community forums. *The Extractive Industries and Society* 2(4): 802–812.

Turton DJ (2017) Legal determinations, geography and justice in Australia's coal seam gas debate. *Natural Resources and Environmental Justice: Australian Perspectives* 1(134): 53–67.

Turton DJ (2020) Lawyers in legal geography: Parliamentary submissions and coal seam gas in Australia. In: O'Donnell T, Robinson DF and Gillespie J (eds) *Legal Geography: Perspectives and Methods*. Routledge, London and New York.

Valverde M (2015) *Chronotopes of law: Jurisdiction, Scale and Governance*. Routledge, London.

Walker G (2012) *Environmental Justice: Concepts, Evidence and Politics.* Routledge, London and New York.

Washington H, Chapron G, Kopnina H, Curry P, Gray J and Piccolo JJ (2018) Foregrounding ecojustice in conservation. *Biological Conservation* 228: 367–374.

CHAPTER 4

World Heritage: Protecting the World's 'Beautiful Places'

Abstract This chapter is about one global protected area type: World Heritage properties. The global environmental treaty for the protection of the world's outstanding cultural and natural heritage has received widespread uptake from countries across the world. Thus, the World Heritage Convention is widely considered to be a successful instrument and many, many iconic places from the Great Wall of China to the Great Barrier Reef in Australia are now protected in ways they would not be, but for their World Heritage status. In this chapter, I consider the evolution of the World Heritage ideal and detail the regulatory dimensions that countries are required to implement for their World Heritage properties—which include management plans, boundaries and buffer zones. My focus on these legal obligations is then applied to case study sites from the Asian, Pacific-Oceanic and Australian region.

Keywords World Heritage · Boundaries · Buffer zones · Management plans · Angkor Cambodia · Trang An Landscape Complex Vietnam · Chief Roi Mata's Domain Vanuatu · East Rennell Solomon Islands · Great Barrier Reef Australia · Tasmanian Wilderness World Heritage Area Australia

J. Gillespie, *Protected Areas*,
https://doi.org/10.1007/978-3-030-40502-1_4

4.1 Introduction

This book is concerned with viewing protected area regimes through a legal geography lens. The larger purpose is to consider how we might best respond to accelerating environmental degradation. In doing this we are considering protected area regime formation and implementation and, ultimately, drawing conclusions about the success of this approach. Protected areas remain the most common way to conserve in-situ habitats, species and areas of cultural heritage significance. I argue throughout this book that the existing protected area system can be improved if we focus greater attention on the human and non-human-place-law nexus that is made clear to us through the legal geography lens. In this and the following chapter, I bring that focus to bear on two significant global environmental governance protected area regimes; World Heritage areas and Ramsar Wetlands. In the eyes of some (though this is a partial view), this emphasis takes us from some of the world's most beautiful to some of the world's less beautiful places, perhaps in recognition of Leopold's (1949, p. 96) observation: "(O)ur ability to perceive quality in nature begins, as in art, with the pretty. It expands through successive stages of the beautiful to values as yet uncaptured by language."

World Heritage listings are, arguably, the best-known form of protected area. There are more than 1000 individual properties on the list, 77% of which are designated as cultural sites, 19% as natural, and the remainder as 'mixed' properties of both outstanding cultural and natural value. Many of humanity's greatest cultural achievements, and some of the Earth's most extraordinary natural features, are protected by the international convention that is promulgated through national law and operationalised at the local scale. From natural sites such as the Great Barrier Reef to cultural sites such as Angkor, a legal geography approach to the study of these outstanding sites gives us place-based informed answers to management challenges.

4.2 Towards World Heritage Protection: A Short History

From the preceding chapter we know that, historically, the rational for the establishment of protected areas is complex and has a very long history. In this book the focus is on the evolution of the protected area system in its modern Western form. In this setting, the rationale for a protected area approach, while complex, can be seen to be a reaction to environmental degradation associated with the industrial revolution and the 'Great Acceleration' of the mid-twentieth century. Additionally, the protected area concept emerged from an era in which dedicated hunting grounds were popular among the elite. In this era the protected area was, undoubtedly, a response to a growing tourism market. While the instinct for preservation has deep roots, the international, co-operative approach to conservation and preservation, best represented by the World Heritage list, is said to have emerged in the late nineteenth and early-mid twentieth century, during a prolonged period of massive global political and economic change (Hall 2016). The institutional infrastructure for cooperation was formalised with the creation of UNESCO, after the roiling chaos of the Second World War (Duedahl 2016).

In the modern era there have been multiple processes at work for assigning protected area status to geographic areas. Some of the key international turning points in promoting protected areas have coincided with the establishment of global environmental organisations, including the International Union for the Conservation of Nature (IUCN) in 1948, and the establishment of the World Wildlife Fund (WWF) in 1961. Another pivotal moment was the United Nations Conference on the Human Environment (otherwise known as the Stockholm Conference producing the Stockholm Declaration), and the establishment of the United Nations Environmental Program, both in 1972. Indeed, the early 1970s was a key time in international environmental policy formulation. One of the key reasons for this was the emergence of one key international environmental convention, the *Convention Concerning the Protection of the World Cultural and Natural Heritage 1972*, otherwise known as the World Heritage Convention.

The World Heritage Convention is widely considered to be an example of a successful global environmental policy. The Convention has almost universal appeal, with 193 parties, even if its utopian founding principles—created by waning colonial powers in the penumbra of war—have been calcified over time by political interests (Meskell 2018). The antecedents of the Convention grew from the success of a dedicated international heritage protection movement which emerged in the aftermath of the damage on cities and places in Europe after World War I and II. Both state and non-state actors were mobilised to 'rescue' famous places from violence and conflict (Logan 2018). The most famous example, illustrative of UNESCO's early mission, is the rescue of the Nubian Monuments (rock temples in Nubia), southern Egypt. In the 1950s the Egyptian government decided to build the Aswan (High) Dam as an embankment dam across the Nile River. The rationale was to control excessive river flooding and to capture water for hydropower electricity generation. These hopes were largely achieved (Hassan 2007). However, the Dam proposal meant a number of significant archaeological sites were threatened by the damming of the river. Amongst the most significant of these were the Nubian Monuments from Abu Simbel to Philae.

The efforts to save the Nubian Monuments are widely considered to be a precursor event towards the establishment of the World Heritage Convention. The salvage campaign began at the behest of the Egyptian and Sudanese governments to relocate the monuments. Governments from across the globe committed tens of millions of dollars in an unprecedented international effort to save the monuments. Between 1964 and 1968 the temples were dissected, labelled and relocated away from the flood zone under the control of UNESCO. The success of the Nubian Monuments project triggered other campaigns—including efforts to save Venice, Italy (1960s) and Borobudur temple compound in Indonesia (1970s). During this era UNESCO teamed with the International Council on Monuments and Sites (ICOMOS) and prepared a draft global agreement to protect the world's cultural heritage (Cameron and Rössler 2011, 2016). At the same time a similar call was made for global efforts to protect natural heritage, through the auspices of the IUCN. Together, ICOMOS, UNESCO and the IUCN drafted text for what was to become the *Convention Concerning the Protection of the World Cultural and Natural Heritage 1972.*

4.3 The World Heritage Convention 1972

In November 1972, in its seventeenth session, the General Conference of UNESCO adopted the *Convention Concerning the Protection of the World Cultural and Natural Heritage.* The text was opened for signature on 16 November 1971 and entered into force 17 December 1975. The Convention brings together a global regime for the protection of both cultural and natural heritage. Article 1 of the Convention defines cultural heritage as a property comprised of 'monuments', 'groups of buildings' or 'sites', whilst Article 2, defines natural heritage as comprised of 'natural features', 'geological and physiographical formations' or 'natural sites'. At present there are 193 parties to the Convention and, as previously mentioned, it is met with such universal acceptance amongst heterogenous nation states that the Convention is often regarded to be one of the most successful global governing instruments (Pocock 1997).

One critical question for protected area analysis is to consider how implementation of the Convention's rhetoric takes place: how the ideals and aspirations are articulated in the Convention and how these translate to a vast array of people and places across the globe. The Convention text, at Article 11, requires state parties to submit to the World Heritage Committee (the organisational arm of the Convention), an inventory of properties for inclusion on the World Heritage List. Based on these inventories the Committee is required to establish the List—based on properties considered to have 'outstanding universal value' (following subsection 2 of Article 11).

Thus, every property nominated and included on the official List must be of 'outstanding universal value'. Outstanding Universal Value (OUV) is the raison d'etre of the Convention. The overarching aim of which is to safeguard places with OUV for present and future generations (Article 4). To understand more about OUV we turn to Paragraph 49 of the Convention's Operational Guidelines:

> Outstanding Universal Value means cultural and/or natural significance which is so exceptional as to transcend national boundaries and to be of common importance for present and future generations of all humanity. As such, the permanent protection of this heritage is of the highest importance to the international community as a whole. (UNESCO 2019a, p. 20)

To gain a World Heritage designation a property must meet the demands of OUV, defined by three criteria. First, the property must meet one or more of the ten criteria where standards are established (for example, the property must "contain superlative natural phenomena or areas of exceptional natural beauty and aesthetic importance" [#7]). Second, the property must match conditions of 'authenticity/integrity' and, third, the property must meet the management practices as set out in the section "Protection and Management" of the Operational Guidelines (paragraphs 96–119, July 2017).

In this regard, the third element—management practices for the property—is of the utmost importance for inclusion on the World Heritage list. The protection and management requirements need policies for (a) adequate regulatory frameworks, (b) to delineate the boundaries of the property, (c) to nominate buffer zones, (d) to create a suitable management system and, lastly, to (e) aim for sustainable use of the property. The mandatory obligations are articulated in Table 4.1 (from UNESCO 2019a).

In assessing these on-the-ground management obligations a legal geography approach comes to the fore. As I argue throughout this book,

Table 4.1 World Heritage Operational Guidelines 2017 regulatory burdens

Management category	*Article*	*Obligation text*
Legislative, regulatory and contractual measures for protection	98	Legislative and regulatory measures at national and local levels should assure the protection of the property from social, economic and other pressures or changes that might negatively impact the Outstanding Universal Value, including the integrity and/or authenticity of the property. States Parties should also assure the full and effective implementation of such measures
Boundaries for effective protection	99	The delineation of boundaries is an essential requirement in the establishment of effective protection of nominated properties Boundaries should be drawn to incorporate all the attributes that convey the Outstanding Universal Value and to ensure the integrity and/or authenticity of the property
	101	For properties (natural) boundaries should reflect the spatial requirements of habitats, species, processes or phenomena that provide the basis for their inscription on the WH List. The boundaries should include sufficient areas immediately adjacent to the area of OUV in order to protect the property's heritage values from direct effect of human encroachment and impacts of resource use outside of the nominated area

(continued)

Table 4.1 (continued)

Management category	*Article*	*Obligation text*
Buffer zones	103	Wherever necessary for the proper protection of the property, an adequate buffer zone should be provided
	104	For the purposes of effective protection of the nominated property, a buffer zone is an area surrounding the nominated property which has complementary legal and/or customary restrictions placed on its use and development to give an added layer of protection to the property. This should include the immediate setting of the nominated property, important views and other areas or attributes that are functionally important as a support to the property and its protection. The area constituting the buffer zone should be determined in each case through appropriate mechanisms. Details on the size, characteristics and authorized uses of a buffer zone, as well as a map indicating the precise boundaries of the property and its buffer zone, should be provided in the nomination

legal geography enables us to evaluate how the human/place/regulation nexus articulates with the rhetoric and mechanistic implementation of environmental protection regimes, in this case, the World Heritage Convention. By enabling us to think carefully about the implications of operationalising law in different places, this perspective can facilitate more effective implementation and, ultimately, act to make environmental and cultural heritage protection more effective.

4.4 Regional Context

There are currently over 1100 world heritage listed properties throughout the globe. UNESCO clusters World Heritage properties according to regions, and the "Asia and the Pacific" category, extending from

Afghanistan to Vietnam, includes more than 250 World Heritage sites. In this book I am particularly focussed the Australasian region, as set out in Table 4.2.

Within the 19 countries listed in Table 4.2 there are 74 sites. Australia (20 sites), Indonesia (9), the Philippines (6) and Vietnam (8) have the greater number of World Heritage properties. Despite the size and diversity of the region, it remains relatively under-represented in terms of the number of properties inscribed and in terms of governance (Meskell et al. 2015).

4.5 Southeast Asia: Competing Pressures

We begin our focus on World Heritage properties situated within mainland Southeast Asia. This is not a comprehensive coverage, but selected case studies are used to draw out the key elements of a Legal Geography approach.

4.5.1 Angkor, Cambodia

In this section, I draw upon my own research work based around the World Heritage site of Angkor, Cambodia. The site was inscribed on the List of World Heritage in Danger in 1992 (Decision 16COMEA) and provides an excellent illustration of the problems that can arise in translating and implementing global obligations into and onto pre-existing populated landscapes. Our focus here is the regulatory obligations to fix boundaries and buffer zones for World Heritage properties, as drawn from Article 5 of the Convention, which requires that 'adequate protection mechanisms' are established through the Operational Guidelines, paragraphs 98–104, as previously identified (see Gillespie 2013 for a discussion of wording of the Convention vis-à-vis boundaries of World Heritage sites).

Angkor is the name given to a group of monuments, constructed between the ninth and thirteenth centuries A.D., in central Cambodia. It was the capital district for the Khmer Empire, became the largest preindustrial city on Earth, and is one of the largest and most important archaeological sites (Coe and Evans 2018). World Heritage boundary and buffer management obligations at Angkor are manifest in a zoned planning approach represented in the "Zoning and Environmental Management Plan" (ZEMP) for Angkor. The ZEMP articulates five planning

Table 4.2 World Heritage properties (all types) in the Australasia-Oceania region

Country	*Property*
Australia	Great Barrier Reef
	Kakadu National Park
	Willandra Lakes Region
	Lord Howe Island Group
	Tasmanian Wilderness
	Gondwana Rainforests of Australia
	Uluru Kata Tjuta National Park
	Wet Tropics of Queensland
	Shark Bay, Western Australia
	Fraser Island
	Australian Fossil Mammal Sites (Riversleigh/Naracoorte)
	Heard and McDonald Islands
	Macquarie Island
	Greater Blue Mountains Area
	Purnululu National Park
	Royal Exhibition Building and Carlton Gardens
	Sydney Opera House
	Australian Convict Sites
	Ningaloo Coast
	Budj Bim Cultural Landscape
Cambodia	Angkor
	Temple of Preah Vihear
	Temple Zone of Sambor Prei Kuk, Archaeological Site of Ancient Ishanapura
Fiji	Levuka Historical Port Town
Indonesia	Borobudur Temple Compounds
	Komodo National Park
	Prambanan Temple Compounds
	Ujung Kulon National Park
	Sangiran Early Man Site
	Lorentz National Park
	Tropical Rainforest of Sumatra
	Cultural Landscape of Bali Provence: The Subak System as a Manifestation of the Tri Hita Karana Philosophy
	Ombilin Coal Mining Heritage of Sawahlunto

(continued)

Table 4.2 (continued)

Country	*Property*
Kiribati	Phoenix Islands Protected Area
Lao People's Democratic Republic	Town of Luang Prabang
	Vat Phou and Associated Ancient Settlements within the Champasak Cultural Landscape
	Megalithic Jar Sites in Xiengkhuang—Plain of Jars
Malaysia	Gunung Mulu National Park
	Kinabalu Park
	Melaka and George Town, Historic Cities of the Straits of Malacca
	Archaeological Heritage of the Lenggong Valley
Marshall Islands	Bikini Atoll Nuclear Test Site
Micronesia (Federated States of)	Nan Madol: Ceremonial Centre of Eastern Micronesia
Myanmar	Pyu Ancient Cities
	Bagan
New Zealand	Te Wahipounamu—South West New Zealand
	Tongariro National Park
	New Zealand Sub-Antarctic Islands
Palau	Rock Islands Southern Lagoon
Papua New Guinea	Kuk Early Agricultural Site
Philippines	Baroque Churches of the Philippines
	Tubbataha Reefs Natural Park
	Rice Terraces of the Philippine Cordilleras
	Historic City of Vigan
	Puerto-Princesa Subterranean River National Park
	Mount Hamiguitan Range Wildlife Sanctuary
Singapore	Singapore Botanic Gardens
Solomon Islands	East Rennell
Thailand	Historic City of Ayutthaya
	Historical Town of Sukhothai and Associated Historic Towns
	Thungyai-Huai Kha Khaeng Wildlife Sanctuaries
	Ban Chiang Archaeological Site
	Dong Phayayen-Khao Yai Forest Complex

(continued)

Table 4.2 (continued)

Country	*Property*
Vanuatu	Chief Roi Mata's Domain
Vietnam	Complex of Hue Monuments
	Ha Long Bay
	Hoi An Ancient Town
	My Son Sanctuary
	Phong Nha-Ke Bang National Park
	Central Sector of the Imperial Citadel of Thang Long—Hanoi
	Citadel of the Ho Dynasty
	Trang An Landscape Complex

Adapted from UNESCO World Heritage Centre, UNESCO (2019b)

zones which vary in the level of land use restrictions; from Zone 1 which is the most highly protected zone with an encircling "buffer zone" (Zone 2), through to Zone 5 which represents the entire Siem Reap administrative unit (the Province, with an area of more than 10,000 km^2). These zoning regulations have the force of law, by virtue of the 1994 *Royal Decree Establishing Protected Cultural Zones in the Siem Reap/Angkor Region and Guidelines for their Management*. However, the extent to which this zone planning policy has been successful at Angko r is disputed, and numerous problems arise in spatially prescribing restrictions on land use.

One critical concern at Angkor was the issue of boundary uncertainty. The World Heritage listing process for Angkor produced a zoning map that bounded the landscape and imposed restrictive land use regulations on places with pre-existing populations and uses. Incompatibility between existing use, expectations and future hopes arose (Gillespie 2013). There was little by way of consultation with communities, largely due to the unusual circumstances in which the listing came to fruition. At the time, Cambodia was newly emerging from civil conflict and the haste with which Angkor was initially placed on the World Heritage list reflects this. The result was that knowledge among local communities of World Heritage boundaries was limited and localised conditions were not reflected in the finalised local zoning map. As a World Heritage property noted for its outstanding monuments and temples, those communities living among the material heritage were not given a prominent voice. Some communities were relocated, while others had (to them) onerous restrictions imposed upon them which limited their ability to build new homes, sell their land and farm their smallholdings (see Gillespie 2009, 2011, 2012, 2013).

A similar implementation problem arose in relation to the twinned regulation requiring boundary fixing—the requirement for buffer zones to be mapped. At Angkor the ZEMP map designated buffer zones which were incongruous with both the lived-in landscape and the pre-existing administrative/institutional realities. Buffer zones carved up villages and went through various local and district administration areas (Gillespie 2012). The buffer zone was both poorly designed and executed. It became clear that, in the heritage management context of Angkor, a significant challenge existing in creating an effective land management system.

All World Heritage properties are subject to monitoring and reporting requirements (a six-year periodic reporting requirement in accordance with Article 29 of the Convention). Another concern arose at Angkor in this context, where disparate laws and lores give rise to differential expectations regarding land use practice and management. The Convention applies to all places, yet it has had a distinctly Eurocentric flavour (Pfeilstetter 2017). Today, the need to account for different perspectives in protecting World Heritage places is *de rigueur* and regarded as best practice (Taylor 2009). The difficulties in translating and implementing a global governance regime to take account of different values and perspectives at localised levels was also evident at Angkor. It became apparent that there was a greater need to account for legally plural arrangements, where there exists a hybrid system of governance with simultaneous localised authority based on customary practices mixed with formalised, State-based regulations, in order to avoid tension in policy outcomes (Gillespie 2011).

A legal geographer's focus on the idiosyncrasies of place and the contingent nature of interaction between people, the spaces they occupy, and the structure of law enables us to improve our understanding of protection regimes. Legal geography's focus on documenting an empirical understanding of these place-based interactions enables it to step outside of assumptions about how international protection mechanisms *should* work at the site level toward a clear-eyed assessment of how they *do* (or do not) work. In the case of Angkor, the legal geography lens reveals profound tensions and injustices that arise from the unalloyed imposition of regulatory regimes and their attendant spatial patterns of exclusion and control.

4.5.2 *Trang An Landscape Complex, Vietnam*

Located on the Red River Delta in Northern Vietnam, the Trang An Landscape Complex was designed as a World Heritage property in 2014. This property is classified as a mixed cultural and natural property—the first of its kind in Southeast Asia—being both a globally-significant karst (limestone) landscape and a site of human occupation for over 30,000 years. Covering more than 6000 hectares, the property is composed of the Hoa Lu Ancient Capital, the Trang An-Tam Coc-Bich Dong Scenic Landscape, and the Hoa Lu Special-use Forest. The buffer zone also covers approximately 6000 hectares. The property contains residential populations, although it is largely rural in character. Paragraph 77 of operational guidelines 2011 sets out 10 criteria for assessment of the OUV. Trang An was nominated pursuant to criteria 5, 7 and 8. These criteria are as follows:

> (v) be an outstanding example of a traditional human settlement, land-use, or sea-use which is representative of a culture (or cultures), or human interaction with the environment especially when it has become vulnerable under the impact of irreversible change;
> (vii) contain superlative natural phenomena or areas of exceptional natural beauty and aesthetic importance;
> (viii) be outstanding examples representing major stages of earth's history, including the record of life, significant on-going geological processes in the development of landforms, or significant geomorphic or physiographic features.

Trang An is under the control of the Ninh Binh Provincial People's Committee, and is entirely owned by the Vietnamese government. The property is managed by the Trang An Landscape Complex Management Board. There is a legally binding management plan for the property that establishes various zones. Despite the successful World Heritage listing, this nomination of this property was not straight-forward. In an IUCN Evaluation of the World Heritage nomination (April 2014) the final recommendation suggested deferring the nomination of the property to:

> … allow the State Party to: a) prepare a revised World Heritage nomination with a boundary that better reflects the areas and attributes of possible Outstanding Universal Value and an appropriate surrounding buffer zone; b) ensure adequate legal protection for the revised nomination including the designation of any areas within the property as protected areas; c) prepare a revised and upgraded Management Plan and Zoning Plan, that recognises the Outstanding Universal Value of the property and ensures that the protection is aligned and integrated into provincial planning. (IUCN 2014, p. 95)

The IUCN commentary is insightful. The regulatory and administrative arrangements for the property gave rise to concerns. The report notes that there are several developments (e.g. new roads, tunnels, car parks, infills and housing developments) that do not warrant inclusion within the nominated World Heritage property. Regarding site boundaries, the assessment concludes that there are concerns with the placement of the nomination boundary and that the configuration of the boundary (and associated buffer zones) needed re-consideration. The IUCN report also indicated that there were a number of inaccuracies in the mapped zoning system whereby existing land use activities conflict with the proposed rules.

As at Angkor in neighbouring Cambodia, the listing of Trang An as a World Heritage property had diverse consequences for local communities. While eco-tourism linked to the property brings new opportunities for local people, Hoang and Pulliat (2019) found "lesser-known struggles and tensions in the community when local residents found themselves in the midst of changes in land use, livelihood, access to resources, and power" (p. 189). The overprinting of international conservation ideals on landscapes filled with normative regulations and long-established practices and patterns of use creates profound tensions and injustices, real or perceived. Paying careful attention to these concerns, and recording them using robust and ethical methods, is critically important. From a legal geographical perspective, the ability to make visible the cross-scale interactions between international environmental obligations and localized laws and customs can improve the design and execution of World Heritage properties, enabling UNESCO to better reflect its founding principles and secure a more just and sustainable management regime for the people that live among the world's natural and cultural heritage.

4.6 The Oceania Region: Invisible Lore's Influence

The practice of Legal Geography captures complexity in governance arrangements that goes beyond law or institutional analysis. Recognising World Heritage governance as co-constituted between formal laws, custom and underlying people-place connections that may have evolved over millennia is a transformative perspective, and one with real significance for good environmental governance. As we have seen, the more than 1000 World Heritage properties are incredibly diverse. This diversity, in of itself, throws up challenges to 'top down' regulatory approaches, forcing them to pay more attention to the idiosyncrasies of place and material. The legal geography perspective is well suited to making these otherwise intangible customary land use practices and human-environment materiality's 'visible'. This is true across all types of World Heritage properties, and for different approaches to their regulation, management and ownership. Customary governance of World Heritage properties is particularly well developed in the Pacific region, and sites in the region were the first in the world to be inscribed on the list that met the governance requirements of the Convention through customary ownership regimes. Notionally, customary arrangements should enable law to be more reactive to people-place dynamics and to interdigitate customary and contemporary practices and patterns of use with formal law and regulation. Certainly, this view is strongly and explicitly developed in UNESCO's Pacific World Heritage Action Plan 2016–2020 (UNESCO 2015). However, customary arrangements in the Pacific also reflect 'chronically weak' centralised state control (Trau et al. 2014, p. 91). The success or otherwise of customary arrangements in the Pacific, then, represents an important test case for the legal geography project.

4.6.1 Chief Roi Mata's Domain, Vanuatu

Chief Roi Mata's Domain is located within the archipelago of Vanuatu, a South Pacific Ocean nation comprised of 83 islands. The site was placed on the World Heritage List in 2008 and is located on the islands of Efate, Lelepa and Artok and the sea between them, in Central Vanuatu. This is the only World Heritage property in Vanuatu. The site represents the last Roi Mata domain. Roi Mata (paramount chief) lived around 1600 AD but his legacy in moral teachings and social reforms lives in oral traditions practiced throughout the Pacific to this day. This World Heritage property is part of his former domain (his home, death location and a mass burial site).

This site is managed under customary tenure arrangements and, because of this, the execution of legal obligations set out in the Convention in this localised setting has been the subject of critical review (Trau et al. 2014). In particular, the implementation of the buffer zone obligation has been thoroughly assessed. In their work Trau and co-workers found that the idea of a buffer zone as a zone of transition to protect the property formulated within the Convention and Operational Guidelines took a distinctively local form, the 'bafa zon'. The authors record reactions to the concept and document the translation process from global obligations to localised enactments. They document contestation, conflict and negotiation from different stakeholder perspectives—providing evidence that the execution of an (idealised) regulation morphs and twists according to competing values. They clearly demonstrate that, far from being inert or homogenous, local communities exert influence, contest and re-shape this fundamental protected area idea.

In this particular case, an evolving sense of *individual* customary ownership of land within the World Heritage buffer zone has eroded the authority of the Council of Chiefs and led to a patchwork of customary ownership and lease arrangements with 'outsiders' that has significantly weakened the capacity for customary governance to sustainable manage the World Heritage site (Wilson et al. 2011).

4.6.2 East Rennell, Solomon Islands

East Rennell in the Solomon Islands was inscribed on to the World Heritage List in 1998 (Decision 22 COM VIIIA.1) and is a natural property being of OUV in relation to its island biogeography. East Rennell forms part of Rennell Island, the southernmost island of the Solomon Island group. It is the largest raised coral atoll in the world and hosts the largest lake in the insular Pacific (Lake Tegano). Importantly, land management at East Rennell is subject to customary tenure where inheritance, allocation of use rights, dispute settlement and record-keeping are managed through social systems. This was the first natural property to be added to the list whilst under a customary land tenure regime. At the time of listing the advisory body's report (the IUCN Technical Evaluation of the nomination), noted that customary land tenure could prevent inscription on the list if it proved prohibitive to a national legislative response, as a successful listing would require (UNESCO 1998). Nonetheless, the property was inscribed on the list and: "(t)his case established an important standard and precedent in relation to the acceptance of customary law and management as a sufficient basis for the management and long term protection of natural World Heritage properties" (UNESCO 2007, p. 33). This may be so, but there has been no comprehensive empirical work to evaluate the success or otherwise of land management practice since the site was inscribed, and the putative value of East Rennell as a precedent for alternative land tenure arrangements has not been tested empirically (Price 2018).

The importance in recognising customary forms of governance as a legitimate alternative to state-based governance is an important step forward in World Heritage management and reflects changing perceptions with regard to local communities and protected area management more widely. While this development in implementation of the Convention is lauded, Smith (2011) presents a useful assessment of the problems associated with this approach. The essential dilemma was that the expectations of customary owners that the listing would reward them was not met. While civil unrest in the Solomon Islands from the time of listing in 1998 through to 2005 caused problems as government processes and communications broke down, Smith asserts that "... the outcome not primarily of the civil strife but of the implementation of the World Heritage Convention in the Solomon Islands" (at p. 596). She documents how the community expectations relating to improvements to livelihoods (especially eco-tourism) were not met because the motives for listing were not to aid protection of the cultural landscape (this was taking place anyway) but by a need to exploit the "'beauty pageant' potential of the World Heritage list" (p. 605). Smith documents how communication and misunderstanding persisted between global expectations and local needs.

In the Pacific region, the agency given to communities and individuals through relatively weak State power and constitutionally-recognised customary ownership arrangements exposes World Heritage norms to diverse, sometimes contradictory and highly-dynamic social change. Far from displacing or overwhelming locality, in the Pacific region at least, World Heritage norms are being forced through a customary filter in what Trau and colleagues describe as 'glocalization'—the translation of global ideas through local perspectives (Trau et al. 2014). This situation does not sit easily with international (effectively Western) expectations of heritage governance and requires locally relevant decision making and conceptions of internationally expected protection mechanisms (such as buffer zones) (Trau et al. 2014) to be seriously recognised and adequately supported by UNESCO. In addition, in the absence of tangible benefits from World Heritage listing for many customary owners, efforts to engage local people in protection should be framed by sustainable development mechanisms that will increase and maintain standards of living for local people (Price 2018)

4.7 Australia: Challenges Within a Federal System

Unlike the situation in the neighbouring Pacific Region, World Heritage properties located within Australia are subject to a clearly structured regulatory hierarchy and strong centralised government control. As a signatory to the Convention, Australia implements her obligations to protect designated properties through national legislation, the *Environmental Protection and Biodiversity Conservation Act 1999* (C'th). World Heritage properties are a 'matter of national environmental significance' under this act so any adverse impacts must meet the national protection standards. However, Australia has a federalised governing system in which the national government (i.e. federal or commonwealth government) must work in cooperation with each of the State (and Territory) governments. For the regulation and ultimately the protection of World Heritage properties this cooperative federalism has been, and continues to be, vexatious.

4.7.1 *The Tasmanian Wilderness World Heritage Area*

The lead up to, and listing of, the Tasmanian Wilderness World Heritage Area (TWWHA) illustrates the challenges of cooperative federalism well. Covering an area of more than 16,000 km^2, the Tasmanian Wilderness incorporates remnants of ancient Gondwanan cool-temperate ecosystems and is recognised as globally significant centre for plant diversity. The rugged, glaciated landscapes have also been home to people for millennia, and the protected area is rich in cultural history.

In 1978 the Tasmanian Hydro-Electric Commission, an agency of the Tasmanian state government, proposed to construct the Franklin Dam, sometimes known as the Gordon-below-Franklin Dam, on the Gordon River in south-western Tasmania. This dam would flood significant parts of the Franklin River and its surrounds. In 1982 the Tasmanian government passed laws allowing the dam to proceed and the Commission began preliminary construction works. However, at the same time, the site (including the river and surrounds) was part of a World Heritage nomination (1981) that was subsequently successful with a full World Heritage listing the following year.

In 1983 an Australian federal election saw the incumbent conservative government succeeded by a Labor party (ALP) government led by Bob Hawke. During this election campaign the damming of the Franklin River was a significant issue, garnering widespread media and activist attention. So much so, in fact, that it became the first environmental campaign to attain a global profile and launched Australia's only 'green' political party (Hutchins and Lester 2006).

The incoming ALP government's commitments to stop the damming of the Franklin was considered to be a pivotal reason behind the new federal government's electoral victory. True to their promise, the new government passed into federal law the *World Heritage Properties Conservation Act 1983* (C'th) and, together with the *National Parks and Wildlife Conservation Act 1975* (C'th), were able to stop construction works (clearing, excavation) for the dam within the newly declared TWWHA.

However, the Tasmanian government challenged the federal government's power to stop the dam in constitutional challenge before the High Court of Australia. Simply, the Tasmanian government argued that the power of the Australian government to prevent the dam was *ultra vires* (outside its power). In a landmark and highly influential case, *Commonwealth* v *Tasmania*, known as the *Tasmanian Dams Case*, the High Court of Australia in a majority decision ruled that the Commonwealth government did have the power to prevent the damming by virtue of its powers under the *Commonwealth of Australia Constitution Act 1900s* 51(xxix) external affairs provision (Rothwell 1993). This was, and remains, the most famous and influential environmental law case in Australian history (Mercer 1985).

Commonwealth jurisdiction for World Heritage properties is well established. As previously mentioned, through the provisions of the *Environmental Protection and Biodiversity Conservation Act 1999* (C'th), the federal government is obligated to protect World Heritage properties located in Australia. How this is executed or operationalised vis-à-vis state governments, however, remains contested. To demonstrate this contestation, we turn to more recent TWWHA controversies (Boer 1992). In

2013 the World Heritage Committee considered and voted for a proposal to extend the boundaries of the TWWHA. More than 170,000 hectares were added to the existing World Heritage area. The extension was in part motivated by the Tasmanian Forest Agreement 'peace deal'. The 'peace deal'—so called to the signal the end of the decades-long conflict over harvesting native forests in Tasmania—officially came into being under the auspices of the Tasmanian Forest Agreement. The peace, as it transpired, was fragile and the World Heritage boundary modification became a highly contentious issue (Kirkpatrick 1998; Lindenmayer 2018). For some these boundary modifications had potential adverse impacts. One local farmer, described by media as holding a fifth generation family farm, was quoted as saying: "I don't know whether there'll be buffer zones imposed, I need to manage the wildlife off that area, I've got patches I intend to log in the future and the implications of all that are just unknown …" (Bryan 2014). A Senate Inquiry into the boundary modification was initiated by the new conservative federal government in 2014. The Inquiry recommended, inter alia, that the boundary modification be formally withdrawn. The federal government subsequently took this boundary modification (to change the boundaries back again) proposal to the World Heritage Committee. Meeting in Doha in 2014 the World Heritage Committee were not amenable to another re-alignment of the boundaries. Portugal's representative on the Committee described Australia's attempt to justify the change the boundaries as "feeble". The Committee unanimously rejected the proposal in less than 10 minutes of deliberation (Mathiesen 2014). The boundary modification was not approved. A 2015 World Heritage Committee Reactive Mission report found:

> Put simply, the 2013 Minor Boundary Modification (MBM) added land to the TWWHA, which is highly valued by some stakeholders for its conservation values, and with some areas targeted for their timber resources by others. The 2013 MBM excluded access to timber in the added area, which was understood to be an integral part of the state-wide Tasmanian Forests Agreement Act 2013 (TFA) at the time. The TFA was described to the mission as a negotiated effort to balance commercial logging and forest conservation in Tasmania. Some stakeholders consulted by the mission referred to the TFA as a successful "peace deal", while others, on both sides of the argument, expressed more critical views. The exclusion of logging in the additions to the property was an explicit objective brought forward by the State Party in its 2013 MBM proposal. Following political changes,

> the State Party submitted a further MBM proposal one year later, which would have partially reversed the 2013 MBM. The World Heritage Committee, however, did not approve the proposal (Decision 38COM 8B.47, Qatar, 2014). (Jaeger and Sand 2015, p. 11)

This complexity of the regulation and management of this large and complex landscape is exacerbated by the federal and state decision-making matrix and a range of stakeholder, land use and tenure types (Rogers and Mackey 2015). A legal geography lens enables us to consider the governance dimensions of this issue with a focus that establishes the importance of the regulatory frame and the structural rigidity of law's hierarchies as a playing a pivotal role in shaping how these places are conserved.

Place-based observation through the eyes of local communities provides the narrative for informed engagement and enhances decision-making protocols (Della Bosca and Gillespie 2019; Davey and Gillespie 2014). Capturing information from all stakeholders, including indigenous perspectives, is critical to this task. In the case of the TWWHA, the World Heritage Committee has been concerned to ensure that the cultural heritage values of the property were also properly accounted for and conserved. However, synthesising data on the indigenous cultural heritage sites are problematic—due in large part to a heterogenous indigenous community that differ about how to appropriately capture (survey) the World Heritage site. Understanding divergent stakeholder perspectives through methodically recording and describing people-place connections is a challenge but is part of the geographer's toolkit (see Logan 2013) and can become part of the *legal* geographer's toolkit (see O'Donnell et al. 2020).

This toing and froing with boundaries increasing and decreasing in a contested political setting creates lived-in problems. Just as in the situation described earlier surrounding the land use classifications for Angkor in Cambodia, local communities in and around the TWWHA ask these same questions: Where is the boundary? Where is the buffer boundary? What rules apply where? Who will enforce the rules and what are the penalties for failing to comply? Critically, they all also ask: How will this listing impact my livelihood? We see a landscape which is shaped by laws and decision making crafted at a global scale—without these regulations these places would not be conserved for generations to come. We can read the legal regime in this landscape—it is encoded on papers but also on our rivers and mountains. Understanding the complexities in these regulatory layers lies at the heart of legal geographical inquiry. Choosing the scale to understand environmental problematisation is consequential.

4.8 The Great Barrier Reef World Heritage Area

Tensions in the state and federal mandates for environmental decision making vis-à-vis World Heritage properties are also evident in any consideration of the efficacy in the management of another iconic Australian World Heritage property, the Great Barrier Reef (GBR), Queensland. The Great Barrier Reef World Heritage Area was placed on the World Heritage List in 1981. Recognised as a site of OUV the GBR meets the natural criteria for inclusion on the World Heritage List. The reef extends over 2300 kilometres along the north-eastern Australian coastline and covers approximately 348,000 km^2. The reef represents the most extensive coral ecosystem on the planet (Grech et al. 2016).

As with the Tasmanian Wilderness, the GBR is protected through the provisions of the Convention which Australia meets through federal statutes; the *Environmental Protection and Biodiversity Conservation Act 1999* (C'th) and, specifically, *Great Barrier Reef Marine Park Act 1975* (C'th). While the federal government has the regulatory power to manage the reef, due to the exigences of cooperative federalism and environmental decision making in Australia, both state and local governments continue to play a role in managing the reef. As the GBR runs adjacent to the Queensland coast, the state of Queensland plays an important role in regulating the reef (Day and Dobbs 2013).

Environmental decision-making in Australia is subject to three tiers of government intervention, (1) federal (Commonwealth), (2) states/territories and (3) local government. Upon federation the Commonwealth *Constitution* did not dedicate environmental decision-making powers to the newly formed federal government, instead these powers remained residually with the respective State governments. However, some environmental decision-making functions today rest with the federal government due to the national government's obligations under the Constitution's external affairs power. This decision-making power was confirmed in the *Commonwealth* v *Tasmania*, or Tassie Dams Case, discussed in the previous section.

The federal and state governments do work cooperatively to manage the reef through intergovernmental agreements to coordinate and implement policy. Yet, because of the complexity within the decision-making matrix, clear responsibility pathways are often obscured. For example, the Queensland government retains regulatory authority for some coastal environs—including areas between the low and high-water marks, internal waters and some islands. Moreover, the Queensland government has the authority to approve coastal developments—including port developments. Concern about the impact of port developments on the Reef have emerged in the past decade. In particular, the issue of expanding port developments proximal to the GBR has been the subject of World Heritage Committee investigation into the state of conservation of the reef, including any impact on the reef's OUV (Douvere and Badman 2012). The GBR is subjected to regular monitoring to ensure the reef remains on the World Heritage List and is not downgraded to the World Heritage List-in-Danger. Since 2012 the World Heritage Committee at its annual meeting has reviewed Advisory Body evaluations for the management approaches of the reef in 2013, 2014, 2015 and 2017 (UNESCO 2019c). The threats to this unique site present clear dangers.

The World Heritage Committee has recognised that improvements to management have taken place, especially with creation and implementation of the Long-Term Sustainability Plan 2050 (LTSP) and the Great Barrier Reef Outlook Report. Yet, in the 2017 World Heritage Committee decision a lack of progress in the implementation of legislation to control land clearing in catchments that discharge into the reef was noted and, more broadly, that more progress would be needed to ensure adequate legislative support for the full implementation of the LTSP outcomes.

Despite considerable efforts to redress environmental degradation of the reef, the overlapping administrative and bureaucratic systems for regulating land use activities remain problematic. Complementary policy coexists under the cooperative approach yet development pressures in the islands, coastal and adjacent hinterlands continue to potentially threaten the integrity of the property. In an era of climate change, adapting management strategies and enhancing cooperative federalism in the environmental decision-making arena are critical to the preservation and conservation of this unique coral reef ecosystem.

4.9 Conclusion

In law we restrict what people can and cannot do: we create criminal and civil penalties when people do not comply with society's rules. For World Heritage protected areas we are endeavouring to value and conserve places across the globe which we deem to have OUV. However, ensuring that the rules for preservation are complied with requires that they resonate with local conditions—this is clearly a challenge in a variety of contexts—from Cambodia to Tasmania.

In geography we look to explain people-place connections. We strive to understand land and water scapes and to explain how each place comes into being and, in a truly interconnected disciplinary exercise, look to the human and more-than-human dimensions of place. There are distinctive spatial components to our analysis. The protected area World Heritage property examples we have explored in this chapter demonstrate how protective laws can legitimise spatial hierarchies. Scalar concerns, especially in thinking through how environmental problems are scaled—come to the fore when we ask, for example, whether local concerns (divergent in themselves) are always trumped by wider governance (regional/state/national/international) demands. Far from being rigid, linear delineations, regulatory processes are inherently complex and dynamic. The legal geography lens helps to unpack these nuances, because of the way regulatory jurisdictions overlap, both spatially and hierarchically.

References

Boer B (1992) World Heritage disputes in Australia. *Journal of Environmental Law and Litigation* 7: 247.

Bryan S (2014) *Tasmanian farmers affected by the forest peace deal in land management limbo*. Australian Broadcasting Corporation, 10 March. Available https://www.abc.net.au/news/2014-03-10/tasmanian-farmers-affected-by-the-forest-peace-deal-in-land-man/5311016. Accessed 25 November 2019.

Cameron C and Rössler M (2011) Voices of the pioneers: UNESCO's World Heritage Convention 1972–2000. *Journal of Cultural Heritage Management and Sustainable Development* 1(1): 42–54.

Cameron C and Rössler M (2016) *Many Voices, One Vision: The Early Years of the World Heritage Convention*. Routledge, London.

Coe MD and Evans D (2018) *Angkor and the Khmer Civilisation*. Thames and Hudson, New York.

Davey M and Gillespie J (2014) The Great Barrier Reef World Heritage Marine Protected Area: Valuing local perspectives in environmental protection. *Australian Geographer* 45(2): 131–145.

Day JC and Dobbs K (2013) Effective governance of a large and complex cross-jurisdictional marine protected area: Australia's Great Barrier Reef. *Marine Policy* 41: 14–24.

Della Bosca H and Gillespie J (2019) The construction of 'local' interest in New South Wales environmental planning processes. *Australian Geographer* 50(1): 49–68.

Douvere F and Badman T (2012) *Report on Reactive Monitoring Mission to Great Barrier Reef (Australia) 6th to 14th March 2012*. UNESCO World Heritage Centre—IUCN. Available http://whc.unesco.org/en/list/154/documents/. Accessed 12 February 2019.

Duedahl P (ed) (2016) *A History of UNESCO: Global Actions and Impacts*. Palgrave Macmillan, London.

Gillespie J (2009) Protecting World Heritage: Regulating ownership and land use at Angkor Archaeological Park, Cambodia. *International Journal of Heritage Studies* 15(4): 338–354.

Gillespie J (2011) Legal pluralism and World Heritage management at Angkor, Cambodia. *Asia Pacific Journal of Environmental Law* 14(1): 1–19.

Gillespie J (2012) Buffering for conservation: The spatial regulation of Angkor's World Heritage landscape. *International Journal of Heritage Studies* 18(2): 194–208.

Gillespie J (2013) World Heritage management: Boundary-making at Angkor Archaeological Park, Cambodia. *Journal of Environmental Planning and Management* 56(2): 286–304.

Grech A, Pressey RL and Day JC (2016) Coal, cumulative impacts, and the Great Barrier Reef. *Conservation Letters* 9(3): 200–207.

Hall M (ed) (2016) *Towards World Heritage: International Origins of the Preservation Movement 1870–1930*. Routledge, London.

Hassan FA (2007) The Aswan High Dam and the international rescue Nubia Campaign. *African Archaeological Review* 24(3–4): 73–94.

Hoang T and Pulliat G (2019) Green for whom? Exploring ecotourism as a climate-adaptation strategy in Trang An, Vietnam. In *Urban Climate Resilience in Southeast Asia*. Springer, Cham, pp. 179–199.

Hutchins B and Lester L (2006) Environmental protest and tap-dancing with the media in the information age. *Media, Culture and Society* 28(3): 433–451.

IUCN (2014) *World Heritage Nomination: IUCN Technical Evaluation Trang An Landscape Complex (Vietnam)—ID No. 1438*. Available https://whc.unesco.org/en/list/1438/documents/. Accessed 10 October 2019.

Jaeger T and Sand C (2015) *Report of the Reactive Monitoring Mission to the Tasmanian Wilderness, Australia*. International Council on Monuments and

Sites, International Union for Conservation of Nature. Available https://whc.unesco.org/en/list/181/documents/. Accessed 12 November 2019.

Kirkpatrick JB (1998) Nature conservation and the Regional Forest Agreement process. *Australian Journal of Environmental Management* 5(1): 31–37.

Leopold A (1949) *A Sand County Almanac and Sketches Here and There.* Oxford University Press, London and New York.

Lindenmayer DB (2018) Flawed forest policy: Flawed Regional Forest Agreements. *Australasian Journal of Environmental Management* 25(3): 258–266.

Logan W (2013) Australia, Indigenous peoples and World Heritage from Kakadu to Cape York: State Party behaviour under the World Heritage Convention. *Journal of Social Archaeology* 13(2): 153–176.

Logan W (2018) Heritage, Sustainable Development and the achievement of peace and security in our world. In: Larsen PB, Logan W (eds) *World Heritage and Sustainable Development: New Directions in World Heritage Management.* Routledge, London, pp. 134–152.

Mathiesen K (2014) UN rejects Australia's 'feeble' bid to strip Tasmanian Forest's heritage status. *The Guardian.* Available https://www.theguardian.com/environment/2014/jun/23/un-rejects-australia-tasmanian-forest-heritage. Accessed 22 November 2019.

Mercer D (1985) Australia's constitution, federalism and the 'Tasmanian Dam Case'. *Political Geography Quarterly* 4(2): 91–110.

Meskell L (2018) *A Future in Ruins: UNESCO, World Heritage, and the Dream of Peace.* Oxford University Press, Oxford.

Meskell L, Liuzza C and Brown N (2015) World Heritage regionalism: UNESCO from Europe to Asia. *International Journal of Cultural Property* 22(4): 437–470.

O'Donnell T, Robinson DF and Gillespie J (eds) (2020) *Legal Geography: Perspectives and Methods.* Routledge, London and New York.

Pfeilstetter R (2017) Culture in heritage on the socio-anthropological notion of culture in current heritage discourses. *Anthropos* 112(2): 609–616.

Price SC (2018) *World Heritage Conservation in the Pacific.* Palgrave Series in Asia and Pacific Studies. Palgrave Macmillan, Singapore.

Pocock D (1997) Some reflections on World Heritage. *Area* 29(3): 260–268.

Rogers N and Mackey B (2015) A wild law perspective on wilderness management: Managing the Tasmanian Wilderness World Heritage Area. *Australasian Journal of Natural Resources Law and Policy* 18(2): 145–165.

Rothwell DR (1993) The High Court and the external affairs power: A consideration of its outer and inner limits. *Adelaide Law Review* 15: 209.

Smith A (2011) East Rennell World Heritage site: Misunderstandings, inconsistencies and opportunities in the implementation of the World Heritage Convention in the Pacific Islands. *International Journal of Heritage Studies* 17(6): 592–607.

Taylor K (2009) Cultural landscapes and Asia: Reconciling international and Southeast Asian regional values. *Landscape Research* 34(1): 7–31.

Trau A, Ballard C and Wilson M (2014) Bafa Zon: Localising World Heritage at Chief Roi Mata's Domain, Vanuatu. *International Journal of Heritage Studies* 20(1): 86–103.

UNESCO (1998) *East Rennell (854) Nomination Dossier*. UNESCO. Available https://whc.unesco.org/uploads/nominations/854.pdf. Accessed 17 September 2019.

UNESCO (2007) *State of Conservation Report East Rennell (Decision31COM 7B.21)*. UNESCO. Available https://whc.unesco.org/archive/2007/whc07-31com-24e.pdf. Accessed 17 September 2019.

UNESCO (2015) *Pacific World Heritage Action Plan 2016–2020*. UNESCO. Available https://whc.unesco.org/document/142213. Accessed 9 September 2019.

UNESCO (2019a) *Operational Guidelines for the Implementation of the World Heritage Convention*. UNESCO. Available https://whc.unesco.org/en/guidelines/. Accessed 28 November 2019.

UNESCO (2019b) *World Heritage List*. UNESCO. Available https://whc.unesco.org/en/list/&order=region. Accessed 14 September 2019.

UNESCO (2019c) *Great Barrier Reef*. UNESCO. Available https://whc.unesco.org/en/list/154/documents/. Accessed 17 October 2019.

Wilson M, Ballard C and Kalotiti D (2011) Chief Roi Mata's Domain: Challenges for a World Heritage property in Vanuatu. *Historic Environment* 23(2): 5–11.

CHAPTER 5

Wetlands: Protecting the World's 'Ugly' Places

Abstract This chapter will examine the implications of another type of protected area regime; Ramsar Convention listed wetlands. The Ramsar governance regime is described and evaluated in this chapter using case study sites from the Australasian region. Through this analysis I argue for a greater reconciliation between the activities of the human communities subject to regulation with biodiversity and wetland conservation targets. Mechanisms by which global legal commitments to wetlands protection need to be reconciled with national and sub-national operational rules and such a reconciliation is now urgently required. Increasingly, an integrative approach to the conservation of biological and cultural diversity prevails. Integration requires that we understand the human dimensions of institutional and governance arrangements surrounding these crucial protected areas. In this chapter we consider case studies from Southeast Asia, the Tonle Sap (Great Lake) of Cambodia, from within Oceania (and we note this region's under-representation) and the lower Murray Darling Lakes, the Coorong and Lake Alexandrina and Albert Wetland Ramsar Site in Australia.

Keywords Ramsar Convention · Wetlands · Tonle Sap Prek Toal Cambodia · Coorong and Lake Alexandrina and Albert Australia · Qoliqoli Cokovata Reef Fiji · Upper Navua Conservation Area Fiji

J. Gillespie, *Protected Areas*,
https://doi.org/10.1007/978-3-030-40502-1_5

5.1 Introduction

In the twentieth century Aldo Leopold made the following observations, published in his famous land ethic book, *A Sand County Almanac* in 1949 (p. 162),

> (T)he marchlands that once sprawled over the prairie from the Illinois to the Athabasca are shrinking northward. Man cannot live by marsh alone, therefore he must needs live marshless. Progress cannot abide that farmland and marshland, wild and tame, exist in mutual toleration and harmony. So with dredge and dyke, tile and torch, we sucked the cornbelt dry, and now the wheatbelt. Blue lake becomes green bog, green bog becomes caked mud, caked mud becomes a wheatfield. Some day my marsh, dyked and pumped, will lie forgotten under the wheat, just as today and yesterday will lie forgotten under the years. Before the last mud-minnow makes his last wiggle in the last pool, the terns will scream goodbye to Clandeboye, the swans will circle skyward in snowy dignity, and the cranes will blow their trumpets in farewell.

Nearly one hundred years later we continue to lament the loss of wetlands habitats around the world, though Leopold's gentle melancholy has been replaced with a distinct note of panic. Wetlands are acutely and disproportionately vulnerable to degradation, fragmentation and loss, and are therefore at the front line of conservation efforts. Wetlands are also among the most biodiverse ecosystems, are central to global biogeochemical cycling, and wetlands within the Australasian region, including mainland Southeast Asia, are among the most diverse examples of these critical ecosystems.

Accordingly, there has long been international recognition of the need to conserve wetlands and, in fact, the international convention for wetland protection, the Ramsar Convention, was the first modern regulatory instrument for the conservation of natural resources at the global scale (Matthews 2013). The Ramsar Convention (hereafter Ramsar, or the Convention) was signed in 1971 but emerged from dedicated international efforts throughout the 1960s. Today, it is the largest network of formally recognised protected areas, with more than 2300 sites that, collectively, cover more than 250 million hectares of land and waterscapes. With 170 contracting parties, the Convention is a wildly successful international instrument for the conservation and 'wise use' of wetlands. Yet despite the success of the Convention as a legal instrument and exemplar of international cooperation, the decline in wetlands and the concomitant decline in biodiversity remain alarmingly high.

5.2 Perceptions and Popular Culture

It is widely understood that wetlands are liminal spaces for humans. They are transitory spaces between earth and water, difficult to define spatially and dynamic temporally. They are difficult to access, difficult to farm and thus impossible to civilise. They are dark, smelly, uncomfortable and unwelcoming. Even their remarkable biodiversity makes them challenging for humans—at once bountiful and full of threatening life. From Bunyan's slough of despond (1678) to Yoda's exile on the swamp world Dagobah (Kasdan and Brackett 1978), wetlands have been viewed, particularly in the West, as antithetical to modernism and representative of primal nature, chaos and moral decay (Giblett 1996). How we perceive wetlands is of real consequence, as our subjective aesthetic valuation of natural environments directly influences how these environments are regulated and protected (Richardson 2018). As I have argued elsewhere (Della Bosca and Gillespie 2019), in order to secure their sustainable conservation, wetlands must be brought in from the periphery of social and political consciousness.

Our psychological and aesthetic unease toward wetlands has deep roots in antiquity, reflecting the inability to cultivate and order these waterscapes. Before agriculture, however, the biological diversity and productivity of wetlands made them cornucopia for humans and other animals and significant attractors in the landscape for mobile or semi-sedentary human groups (Nicholas 1998). Indeed, recent genetic and paleo-environmental evidence suggests that modern humans first emerged from a vast wetland complex in Southern Africa 200,000 years ago, before climate change encouraged human groups to spread across the planet (Chan et al. 2019). We are all, therefore, swamp people.

Wetlands are globally ubiquitous, occurring in any environment where solid earth transitions to water. A formal international definition of wetlands, as provided in Article 1 of the Ramsar Convention, incorporates "marsh, fen, peatland or water, whether natural or artificial, permanent or temporary, with water that is static, flowing, fresh, brackish or salt, including areas of marine water the depth of which at low tide does not exceed six metres". Such a sweeping definition, incorporating large areas of the Earth's surface, might seem deliberately vague, or might even been seen as an ambit claim by the Convention. However, as Anthony Wilson reminds us in his book *Swamp* (Wilson 2018), the formal definition of wetlands is of critical importance, for what we cannot define in law we cannot protect in law. Whether an area is or is not a wetland will determine if it is to be protected or transformed into pasture or drained and given over to urban development. Of course, wetland's highly variable form and frequently ephemeral nature defy tight definition—hence the sweeping definition adopted in the Ramsar Convention.

Wetlands are also a major source of greenhouse gases, particularly methane. The decomposition of organic material in wetlands soils and sediment under anoxic (oxygen limited or oxygen free) conditions produce 109Tg CH_4 globally per year (with considerable uncertainty due to spatially patchy data), the bulk of which comes from tropical wetlands (Bartlett and Harriss 1993). Methane is such a potent greenhouse gas, and wetlands such an important source of it, that the expansion of tropical wetlands in the geological past has been linked with planetary-scale changes in atmospheric chemistry and climate (Ruddiman 2003).

5.3 Losses Are High

Despite the importance of natural wetlands in regulating the earth's systems and maintaining high biological diversity, they are among the most denuded of natural environments and have undergone significant losses, both in terms of area and function, as a result of human activity. Quantifying precisely how great an area of the Earth's wetlands has been lost is critically important and infuriatingly difficult—important because knowing what we have lost or are likely to lose in the future is core to global methane and carbon budgets, and difficult because global historical data are regionally patchy and because wetlands are highly variable in both definition and in reality (Hu et al. 2017; Davidson 2014). Recent research by the United Nations Environment Program reveals that natural wetlands, in particular, declined in extent by 30% between 1970 and 2008 (Dixon et al. 2016). Ramsar's own Global Wetland Outlook (Ramsar Convention on Wetlands 2018) reveals that the extent of wetlands globally decreased by 35% between 1970 and 2015. This recent acceleration in the loss of wetland area is part of a much longer-term trend—as much as 87% of wetlands may have been lost since the start of the eighteenth century (Davidson 2014).

It is clear that wetlands have suffered disproportionately as a result of human activity. This is in part a reflection of our practical, psychological and aesthetic ambivalence to wetlands but also because wetlands are resource-rich and are thus particular targets for modification by humans. Wetlands are loci for the sequestration of sediment and other biogeochemically important materials, largely from river catchments, and are thus highly productive.

So, despite the prevalence of the global commitment to protect wetlands (Ramsar Convention) and clear data on both the importance of and threats to wetland conservation, the loss and degradation of wetlands globally has not been halted (Finlayson 2012). Finlayson labels this as a policy and implementation paradox ('Finlayson's paradox'): without effective national commitments to effectively implementing protection, global commitments—no matter how well subscribed—are potentially undermined. Arguably, it is at this policy implementation stage that the work of legal geographers comes to the fore (see Chapter 3 and Fig. 3.1). In the following sections we will consider the implementation stage of wetlands laws using our three-prong regional approach.

5.4 Wetlands in Southeast Asia: Cambodia's Great Lake (the Tonle Sap)

In this section we consider the protection and management of the Great Lake (in Khmer *Tonle Sap*) of central Cambodia, in southeast Asia. The Tonle Sap lake sits at the end of massive continental drainage network and inherits the cumulative impact of human activity and modification. In this case, the Mekong River—one of the longest rivers on Earth—drains an area of almost 800,000 km^2, stretching from the Tibetan Plateau at an elevation of more than 5000 m to the South China Sea in Vietnam. The Tonle Sap Lake is located in the lower Mekong Basin, some 420 kilometres by river upstream of the Delta front in the South China Sea and, hydrologically at least, should be considered as part of the Mekong's massive delta system, itself the third largest delta in the world (Li et al. 2017). The Tonle Sap is connected to the Mekong River mainstream by the Tonle Sap river, the confluence between the two rivers occurring adjacent to the Cambodian capital of Phnom Penh. During the intense rainfall of the summer monsoon season in mainland Southeast Asia, flooding in the Mekong River mainstream pushes floodwater upstream, backing up floodwater into central Cambodia and inundating the Tonle Sap lake and its surrounding floodplains. During that time, the lake depth increases from an average of less than 1 m to a maximum of more than 9 m. As much as 79 cubic kilometres of flood water is forced into central Cambodia through the connecting river channels and via overland flow, and the lake expands in area several fold to accommodate this influx, inundating a massive area—nearly 11,000 km^2—of wetlands and floodplains (Kummu and Sarkkula 2008).

Amongst the flood water are tiny mineral particles weathered from the rocks of the Mekong's catchment. Approximately 5.1 billion kilograms of sediment is transported into the Tonle Sap lake basin from the Mekong River each year, and only 20% of that sediment is transported back out of the basin once the flood recedes during the dry season (Kummu et al. 2008). The rest of the sediment is retained in the system, and deposited either in the lake basin itself or, more commonly, over the floodplain. This annual dump of suspended sediment drives enormous primary productivity, because nutrients are preferentially adsorbed to clay particles, and provides the basis for one of the most productive wetlands on Earth (Arias et al. 2013). The extensive riparian wetlands and seasonal flood pulse makes the lake an important nursery habitat for migratory fish species from the Mekong River. The annual movement of fish from the Mekong to the Tonle Sap represents one of the largest animal migrations on Earth in terms of the number of individual animals involved (estimated to be several billion), and makes the lake one of the most productive inland fisheries, with annual fish yield from the lake ranging from 180,000 up to 250,000 tonnes per annum, while the *dai* fisheries (stationary bag nets) on the Tonle Sap river capture 12–14,000 tonnes of fish annually (Lamberts 2006; Sarkkula et al. 2005). This extraordinary bounty has made the Tonle Sap lake attractive to humans from antiquity, and the lake and wetlands provide habitat to globally threatened or endangered more-than-human species, particularly water birds, but also mammals, reptiles and fish.

The unique wetland ecology of the Tonle Sap lake and floodplain has attracted several forms of protected area regulation, creating a complex hierarchical matrix. In the Tonle Sap lake basin alone, there are three separate Ramsar designated wetland areas with a total area of over 580 km^2. In addition, these same wetlands were designated as 'core areas' by UNESCO in 1997 as part of their Man and Biosphere Reserve Program, and the Angkor World Heritage Park (Chapter 4) sits on the northern shore of the lake. All of these international obligations are promulgated through national laws and operationalised through a patchwork of responsibilities that are shared between several government ministries (see Gillespie 2018 for more detail). This teetering stack of formal law becomes meaningful at the local level, where 'floating village' communities that live among the wetlands have their own informal regulations and long-established patterns of use.

My own work among these wetland villages indicates that the imposition of international obligations through national law has occurred with little consideration of vernacular practices, lore or custom, and in many cases individuals and communities feel marginalized or disenfranchised from their own waterscapes and traditional lifeways. Communities that are dependent upon fishing in the wetland express frustration and resentment toward the regulations that they rightly perceive to have been imposed by others from somewhere else, creating tension that can facilitate non-compliance with rules designed to protect biodiversity. This tension between formal law and the dynamics of place has had a number of other, collateral impacts on these communities. Existing power structures within wetland communities appear to be strengthened and more deeply entrenched as a result of the imposition of protected area regulation (Gillespie 2018), while the impact of regulation appears to have gendered implications among wetland communities. This was particularly the case in villages adjacent to the Ramsar sites at Prek Toal and Boeung Chhmar, where female informants reported that protected area regulations had restructured their movements through the wetlands and created a sense of personal insecurity as a result (Gillespie and Perry 2019; Perry and Gillespie 2019). Insights informed through the application of a legal geography approach to this critical wetland site has uncovered important disconnections between law's aims and its application. The spectre of Finlayson's paradox looms large here.

5.5 Wetlands in Oceania: Lore's Influence in an Under-Represented Region

The Ramsar Secretariat lists 8 nations with Ramsar wetlands in this western Pacific region: Australia (66), Fiji (2), Kiribati (1), Marshall Islands (2), New Zealand (6), Palau (1), Papua New Guinea (2) and Samoa (2)

(Ramsar country profiles, n.d.). Accordingly, there are a combined total of 82 sites, with Australia having over 80% of these listed wetlands. Of the more than twenty Pacific Island countries and territories, only six host wetlands of international significance as defined by the Ramsar Convention. This is, perhaps, unsurprising given the relatively small cumulative land area across the Pacific region, which is comprised of hundreds of 'high islands' that are normally volcanic in origin, and thousands of limestone or coral 'low islands' (Nunn et al. 2016). In these circumstances, drainage basins are relatively small and their attendant rivers short, with the exception of Papua New Guinea—the largest area of contiguous land area in the region—which supports rivers such as the Fly and Sepik Rivers. Wetland types represented in the Pacific region include floodplains and rivers and streams, lakes, swamps, marshes, mangroves, seagrass beds and coral reefs (SPREP 2011), with coral reefs being the most abundant type as defined by the Ramsar Convention. Despite the limited land area, endemism is very high in the wetlands of the Pacific region, particularly so for freshwater wetlands which are extremely genetically isolated from comparable wetlands on other islands. Biological diversity is also very high, particularly in the west of the region and declining toward the east (Ellison 2009).

Unlike many Ramsar wetland sites elsewhere in the world, many internationally significant wetlands in the Pacific region are under customary ownership and managed on the owner's behalf by community-based management committees, boards or similar. This, in principle, ensures that the complex suite of formal law and regulation created to protect and manage these places should be more sensitive to place dynamics and the complexities of normative lore. Indeed, Ellison (2009) points to a number of successes in locally based management of wetlands across the region, and Ewel (2008) demonstrate excellent ecological management outcomes derived from local management in the context of overarching, regional regulations. Sites such as the Lake Ngardok Nature Reserve in Palau (Ramsar Site Information Service Site 1232, n.d.), for example, are strictly managed and are utilised for education and scientific research only.

Despite this, wetlands are still being degraded across the Pacific region. Reflecting a long history of coastal wetland utilisation, many of the site are utilised by the customary owners for their diverse ecosystem services. In some cases, such as the isolated Namdrik Atoll in the Marshall Islands (Ramsar Site Information Service Site 2072, n.d.), the intensity of that use has placed pressure on the biodiversity of the protected wetlands. This increasing intensity of wetland exploitation reflects growing populations throughout the Pacific region and internal migration within island nations to the capital island. Throughout the region, pressure on coastal fisheries from subsistence and commercial fishing has seen a consistent decline in fish yield and diversity from coastal and shallow marine fisheries, with some exploited species now endangered or driven to local extinction in Fiji (Ellison 2009).

Threats to wetlands in Fiji are various but include run-off pollution derived from development and deforestation in adjacent terrestrial locations. Pollution from adjacent land use has degraded or threatens to degrade the Qoliqoli Cokovata reef off northern Fiji listed on to the *Ramsar* list in 2018 (Ramsar Site Information Service Site 2331, n.d.). Similarly, deforestation for agriculture in the often-steep catchments that support freshwater wetlands has resulting in widespread slope mobilisation and erosion, leading to the rapid filling of coastal wetlands and the pollution of near-shore reefs and sea-grass systems with mobilized sediment. In the Fijian Upper Navua Conservation Area the forests surrounding the gorge, which constitutes the protected wetland, are threatened by the potential of increased logging activities (Ramsar Site Information Service Site 1612, n.d.). The area is owned by local families.

These pressures, in concert with logging for wood or fuel, reclamation or the effects of invasive plant or animal species, present an enormous regulatory challenge for the customary owners of areas adjacent to the protected wetlands and for the States that sign and operationalise international protection regimes. As a consequence, the Secretariat of the Pacific Regional Environment Programme (SPREP) has repeatedly called for deeper integration with local communities, and more effective engagement in developing 'wise-use' strategies that underpin sustainable management of important wetlands. The (now somewhat dated) Regional Wetlands Action Plan for the Pacific Islands calls for a 50% increase in community participation in wetland conservation, in concert with the development of site-specific plans of management in the context of 'adaptive' national legislation (SPREP 2011). Ellison (2009) tells us, as we have seen Southeast Asia, that "[t]op-down declarations of protected areas have resulted in little protection, while those in community-run conservation areas are governed by local committees with wide powers through the village social system" (p. 199). Even in the case of customary ownership of protected wetlands, the disjunction between national legislation and long-established people-place dynamics creates a tension in the day-to-day management and use of wetlands that, in many cases, is resulting in the degradation and loss of critical wetland habitat. This disjunction requires a better understanding of localised dynamics to articulate with national and international wetland protection commitments. Research inflected with a legal geography dimension can achieve this objective. Using the approach identified in Chapter 3, where legal geographers carefully map local conditions against existing and future regulatory burdens and take into account the scalar implications of policy reforms, we can push towards better environmental and social outcomes.

5.6 Wetlands in Australia: The Lower Murray Lake System

Australia was one of the first contracting parties to the Ramsar Convention. Australia has 66 Ramsar listed wetland properties with a total surface area exceeding 83,000 square kilometres. The wetlands are situated across the entire continent, including Tasmania, and are considered to be 'matters of national environmental significance' under the Commonwealth (federal) government's key environmental protection legislation, the *Environmental Protection and Biodiversity Conservation Act 1999* (C'th). Under this legislative status, any development project that might impact the wetlands must meet national standards of protection and mitigate against adverse impacts.

Yet, at the time of writing the fate of one of these wetlands, the Macquarie Marshes, is in doubt. The Macquarie Marshes were recognised as internationally significant and designated as a Ramsar site in 1986, primarily as a result of their critical role in providing habitat for migratory bird species. The Marshes, which cover an area in excess of 2500 km^2 in full flood, support the largest area of reedbeds in south-eastern Australia and is home to a significant redgum woodlands (Ramsar Site Information Service Site 337, n.d.).

In October 2019, the Macquarie Marshes were on fire (Davies 2019). Already heavily degraded by lack of inflow as a result of river regulation and abstraction of river water, primarily for irrigation (Kingsford and Johnson 1998), ongoing drought has created conditions that promote intense fire that, with the additional ambient stress of climatic change, may engender permanent change in the character of the wetland (Kingsford 2011). Nearly a fifth of the wetland area has been burnt. This particular crisis of the Macquarie Marshes wetlands sets the scene for this section. In this part we consider Ramsar wetland protection within the Murray-Darling river basin catchment. I outline how protection is failing our wetlands and focus in on an example from the lower basin, the Coorong and Lakes Alexandrina and Albert Ramsar wetland (Ramsar Site Information Service Site 321, n.d.).

Wetland protection in the Murray Darling River Basin has been described as 'ineffective' (Kingsford et al. 2011). There are more than 30,000 wetlands in the basin, 16 of which are on the Ramsar list of wetlands, and all are said to be degrading rapidly. Frustration has been voiced by a notable conservation biologist working in wetland and river management, Richard Kingsford, who notes that, despite the legislative protection for these systems, there is ongoing degradation (see Kingsford et al. 2011). Here again we see Finlayson's Paradox. Della Bosca and Gillespie (2019) note that what appear to be well-crafted and comprehensive legal protections, crafted by the international community using the best available science, ratified at the national level and operationalised through state law, are being systemically undermined by political expediency and concessions to vested commercial interests.

This is a consistent picture across Australia, and the Murray Darling Basin in particular. The Murray Darling Basin is Australia's largest river basin, stretching over more than 1000 km^2—about the size of Egypt—and represents 14% of Australia's total land area. Over 13 degrees of latitude the basin stretches from sub-tropical to temperate climates, and from the mountains of the Great Dividing Range to the desert of western Queensland and NSW. The basin is also one of Australia's most productive regions, producing well over $20 billion in food and fibre each year—around 30% of the national food supply. Given its important role in the national economy and as the 'food bowl' for Australia, the basin's rivers are heavily regulated. Regulation in the form of wiers, dams and other infrastructure began in earnest in 1922, but accelerated dramatically in the late 1950s (Leblanc et al. 2012). Now, the total public storage capacity within the basin greatly exceeds the average annual discharge in its rivers (Kingsford et al. 2011).

Perhaps the most serious threat to river and wetland health on the basin is the volume of water abstracted directly from the river to support irrigated agriculture. 70% of Australia's irrigated land lies within the bounds of the Murray Darling basin, and the overwhelming majority of water used in the basin—90% or more—goes to irrigation (Grafton et al. 2014). By the turn of the second millennium, the Murray-Darling was in the grip of a severe drought. The strain put on the natural environment by over-allocation of surface water in concert with systemic river regulation was so dramatic that limits were placed on the amount of water that could be taken from the system. This 'sustainable diversion limit', currently set to 10,873 GL/year, is part of a basin-wide water trading system that is intended to secure water for environmental purposes (so called 'environmental flows') (Hart 2016). However, pressure to reduce the amount of water returned to the environment has seen the volume of water returned to the environment progressively lowered. These political and governance issues will be greatly exacerbated by projected climatic change and its implications for water sharing. Invariably, wetlands will suffer disproportionately as a result of a projected decrease in stream flow and more frequent drought events (Pittock and Finlayson 2011).

To some degree, many of the environmental issues surrounding river regulation, water abstraction for industry, and declining water quality, propagate and magnify downstream, such that wetlands at the end of the system tend to be among the most impacted. This is certainly the case for the network of wetlands that mark the terminus of the Murray-Darling River system. Located in South Australia, the lower lakes—formally The Coorong, and Lake Alexandrina and Albert Wetland Ramsar Site (Ramsar Site 321)—is a network of interconnected wetlands, lakes and lagoons, covering an area of 1425 km^2 that marks the transition from the river to the ocean. The extraordinary diversity of wetland types within that area—as many as 23 Ramsar wetland types—creates a mosaic of habitats that support many vulnerable, endangered and critically endangered species. The site's significance in providing habitat to migratory water birds, in particular, was the primary reason for its designation as a Ramsar site in 1985. A count of waterbirds in the lower lakes in 2000 identified an average of 150,000 individual waterbirds (Paton et al. 2009), with a reported maximum of a staggering 329,000 individuals (O'Connor et al. 2012).

The lower lakes today formed during the early Holocene—around 6–8000 years ago—when sea levels stabilised after the global recovery from the most recent ice age. The complexity of lower lakes, however, reflects the fact that these recent landforms are imprinted over similar systems of much greater antiquity (Cann and Murray-Wallace 2012). Large beach-barrier systems were established across the mouth of the Murray River, creating relatively shallow embayments that were variously influenced by marine water ingress through a breach in the dune barrier (the 'Murray Mouth') and by freshwater discharge from the Murray River itself. Lakes Alexandrina and Albert are connected to one another and to the Murray River, and so are fresh to brackish, while the Coorong—a narrow and shallow lagoon that runs parallel to the beach barrier for 150 km—is most strongly influenced by marine water, and so is variously saline to hypersaline. Under natural conditions the three main water bodies were hydraulically connected to each other through a series of narrow channels, and to the Southern Ocean. However, connection between the lakes was severed after the construction of barrages across these narrow channels in 1935–1940 (Webster 2010). The consequence has been a change in water quality and water level with cascading implications for wetland ecology, biogeochemical cycling and hydrological function (Kingsford et al. 2011). The impact of river regulation and water extraction upstream, and changes to the natural connectivity of the lower lakes, have resulted in a widely-noted deterioration in a range of environmental indicators (Kingsford et al. 2009).

Arguably the case of the Murray Darling River basin and its wetlands presents a clear case of regulatory failure. A legal geography approach to re-thinking the efficacy of wetland regulation may provide better, integrated policy-oriented work that is much needed to future-proof the management of this fragile ecosystem. Taking one core idea from legal geography, that laws are more effective and better outcomes secured, if laws are better tailored to place and the ways in which people value place, then the wetlands management should be tailored to take account of and react to localised conditions. Contemporary regulatory failure in the Murray Darling system is in part caused by the legacy of past land and water use—through hard infrastructure like dams and barrages constructed during the twentieth century—that exert a strong inertial force on contemporary management. It is also a failure to articulate overarching regulatory principles and approaches with the expectations of local communities. This was most famously demonstrated by the public burning of printed copies of the Basin Plan in October 2010, in Griffith, NSW, five days after the official launch of the Plan. While this is said to have been carefully stage-managed by irrigators threatened by water allocations for environmental purposes (Alexandra 2017), it is reflective of a broader hostility to the top-down regulatory approach adopted by the Federal Government on this issue and the lack—real or perceived—of consultation with Basin communities. The subsequent frantic back-pedalling by the Basin Authority the following year, changes in senior staff and a commitment to real engagement, as Barry Hart says, probably saved the Plan (Hart 2016). From a legal geographical perspective this is classic regulatory failure—laws created and imposed with little consideration for the idiosyncrasies of place invite non-compliance and result in poor environmental outcomes, in addition to weakening the rule of law in general.

However, regulation of large systems like drainage basins has to be undertaken from a whole-of-system approach. It serves little purpose to carefully regulate water allocations for environmental outcomes in the lower lakes of the Murray Darling system if all the water is being used in the upstream reaches. Managers of large-scale environmental systems must therefore cast a regulatory net over many places, many of which will have vernacular understandings of place that might be quite at odds with other places within the system. In a drainage basin covering an area of more than 1 million square kilometres that is home to more than 2 million people, any whole-of-system regulation or reform is inevitably going to irritate someone. This raises a particular challenge for the legal geography perspective I have suggested herein. How can laws accommodate place in the context of large interrelated systems without degenerating to an unworkably complex mosaic of place-based regulation—a patchwork quilt of law that, while potentially securing better outcomes through compliance at the local level, would be administratively complex at the state and federal level (and would no doubt present fertile fields for new case law)? This challenge is not insurmountable. A legal geography perspective is enabled by the geographer's ability to account for complexity in both spatial and temporal dynamics. The solution is not a one-size-fits-all response. A greater understanding of people-place-law dynamics, which lies at the heart of legal geography scholarship, can aid and not detract from our appreciation of such complexities and help us to craft appropriate regulatory responses.

5.7 Conclusion

There is a consistent pattern of wetland degradation across the planet. Ramsar's most recent Global Wetland Outlook (Ramsar 2018) reveals that approximately 35% of wetlands have been lost globally since 1970, at a rate up to three times that of forest loss. While Ramsar claim that tools such as the Montreux Record—a lists of sites that have undergone or are likely to undergo change in ecological character—can identify wetland degradation and tacitly encourage governments to create or modify policies and practices, Hamman et al. (2019) argue that the process lacks real potency in generating change (the authors note that more than half of the wetlands placed on the lists initially in from 1990, are yet to be taken off). Regardless, whatever the effectiveness of international 'soft-law' agreements or national/state 'hard' law legislation, the consistent picture that emerges through the legal geography lens is the importance of the adaptive, place-based and engaged policy to secure better environmental outcomes in the management of wetlands, and to avoid instances of regulatory failure that results from tension between place and law. Ramsar has long encouraged the active integration of local people, customary owners and indigenous stakeholders in wetland management and conservation, and this remains a key part of the Convention's current strategic plan. As we have seen in this chapter, such as approach is not without its challenges, nor is it a panacea (Lane and Corbett 2005). The legal geography project demands that we move beyond recognising or 'working with' the place-based concerns and practices of local people (indigenous or otherwise) once conservation policy has been set and implemented, by tempering international and state-based law *as it is forged* with the lived realities of the places where those laws will touch the earth.

References

Alexandra J (2017) Risks, uncertainty and climate confusion in the Murray–Darling Basin reforms. *Water Economics and Policy* 3(03): 1650038.

Arias ME, Cochrane TA, Norton D, Killeen TJ and Khon P (2013) The flood pulse as the underlying driver of vegetation in the largest wetland and fishery of the Mekong Basin. *Ambio: A Journal of the Human Environment* 42(7): 864–876.

Bartlett KB and Harriss RC (1993) Review and assessment of methane emissions from wetlands. *Chemosphere* 26(1–4): 261–320.

Bunyan J (1678) *The Pilgrim's Progress*. Reprint, Classics of World Literature, London 1996, 272 pages.

Cann JH and Murray-Wallace CV (2012) Interstadial age (MIS5c) beach-dune barrier deposits in the Coorong Lagoon, South Australia. *Australian Journal of Earth Sciences* 59(8): 1127–1134.

Chan EKF, Timmermann A, Baldi BF, Moore AE, Lyons SSL, Kalsbeek AMF, Petersen DC, Rautenbach H, Fortsch HEA, Bornman MSR and Hayes VM

(2019) Human origins in southern African palaeo-wetland and first migrations. *Nature* 575: 185–189.
Davidson NC (2014) How much wetland has the world lost? Long-term and recent trends in global wetland area. *Marine and Freshwater Research* 65(10): 934–941.
Davies J (2019) *Ramsar-protected Macquarie Marshes wetland on fire with 90pc of crucial reed bed razed.* Australian Broadcasting Commission, 28 October. Available https://www.abc.net.au/news/2019-10-28/macquarie-marshes-on-fire-90pc-reed-bed-razed/11645914. Accessed 24 November 2019.
Della Bosca H and Gillespie J (2019) Bringing the swamp in from the periphery: Australian wetlands as sites of climate resilience and political agency. *Journal of Environmental Planning and Management*. https://doi.org/10.1080/09640568.2019.1679100.
Dixon MJR, Loh J, Davidson NC, Beltrame C, Freeman R and Walpole M (2016) Tracking global change in ecosystem area: The Wetland Extent Trends index. *Biological Conservation* 193: 27–35.
Ellison JC (2009) Wetlands of the Pacific Island region. *Wetlands Ecology and Management* 17: 169–206.
Ewel KC (2008) Mangrove crab (Scylla serrata) populations may sometimes be best managed locally. *Journal of Sea Research* 59(1–2): 114–120.
Finlayson MC (2012) Forty years of wetland conservation and wise use. *Aquatic Conservation: Marine and Freshwater Ecosystems* 22(2): 139–143.
Giblett RJ (1996) *Postmodern Wetlands: Culture, History, Ecology*. Edinburgh University Press, Edinburgh, 268 pages.
Gillespie J (2018) Wetland conservation and legal layering: Managing Cambodia's great lake. *The Geographical Journal* 184(1): 31–40.
Gillespie J and Perry N (2019) Feminist political ecology and legal geography: A case study of the Tonle Sap protected wetlands of Cambodia. *Environment and Planning A: Economy and Space* 51(5): 1089–1105.
Grafton RQ, Pittock J, Williams J, Jiang Q, Possingham H and Quiggin J (2014) Water planning and hydro-climatic change in the Murray-Darling Basin, Australia. *Ambio: A Journal of the Human Environment* 43(8): 1082–1092.
Hamman E, Van Geelen T and Akhtar-Khavari A (2019) Governance tools for the conservation of wetlands: The role of the Montreux Record under the Ramsar Convention 2019s. *Marine and Freshwater Research* 70(11): 1493–1502.
Hart BT (2016) The Australian Murray–Darling basin plan: Challenges in its implementation (part 1). *International Journal of Water Resources Development* 32(6): 819–834.

Hu S, Niu Z, Chen Y, Li L and Zhang H (2017) Global wetlands: Potential distribution, wetland loss, and status. *Science of the Total Environment* 586: 319–327.

Kasdan L and Brackett L (1978) *Star Wars: The Empire Strikes Back Script at IMSDb* [online]. Imsdb.com. Available http://www.imsdb.com/scripts/Star-Wars-The-Empire-Strikes-Back.html. Accessed 1 November 2019.

Kingsford RT (2011) Conservation management of rivers and wetlands under climate change—A synthesis. *Marine and Freshwater Research* 62(3): 217–222.

Kingsford RT and Johnson W (1998) Impact of water diversions on colonially-nesting waterbirds in the Macquarie Marshes of arid Australia. *Colonial Waterbirds* 21: 159–170.

Kingsford RT, Fairweather PG, Geddes MC, Lester RE, Sammut J and Walker KF (2009) *Engineering a crisis in a Ramsar wetland: The Coorong, Lower Lakes and Murray Mouth, Australia.* Australian Wetlands and Rivers Centre, University of New South Wales, Sydney, 56 pages.

Kingsford RT, Walker KF, Lester RE, Young WJ, Fairweather PG, Sammut J and Geddes MC (2011) A Ramsar wetland in crisis—The Coorong, Lower Lakes and Murray Mouth, Australia. *Marine and Freshwater Research* 62(3): 255–265.

Kummu M and Sarkkula J (2008) Impact of the Mekong River flow alteration on the Tonle Sap flood pulse. *Ambio: A Journal of the Human Environment* 37(3): 185–193.

Kummu M, Penny D, Sarkkula J and Koponen J (2008) Sediment: Curse or blessing for Tonle Sap Lake? *Ambio: A Journal of the Human Environment* 37(3): 158–163.

Lamberts D (2006) The Tonle Sap Lake as a productive ecosystem. *International Journal of Water Resources Development* 22(3): 481–495.

Lane MB and Corbett T (2005) The tyranny of localism: Indigenous participation in community-based environmental management. *Journal of Environmental Policy and Planning* 7(2): 141–159.

Leblanc M, Tweed S, Van Dijk A and Timbal B (2012) A review of historic and future hydrological changes in the Murray-Darling Basin. *Global and Planetary Change* 80: 226–246.

Leopold A (1949) *A Sand County Almanac and Sketches Here and There.* Oxford University Press, London and New York.

Li X, Liu JP, Saito Y and Nguyen VL (2017) Recent evolution of the Mekong Delta and the impacts of dams. *Earth-Science Reviews* 175: 1–17.

Matthews GVT (2013) *The Ramsar Convention on Wetlands: Its History and Development.* Ramsar Convention Bureau, Gland, Switzerland, 88 pages.

Nicholas GP (1998) Wetlands and hunter-gatherers: A global perspective. *Current Anthropology* 39(5): 720–731.

Nunn PD, Kumar L, Eliot I and McLean RF (2016) Classifying Pacific islands. *Geoscience Letters* 3(1): 7.

O'Connor J, Rogers D and Pisanu P (2012) *Monitoring the Ramsar Status of the Coorong, Lakes and Murray Mouth: A Case Study Using Birds*. South Australian Department for Environment and Natural Resources, Adelaide, 38 pages.

Paton DC, Rogers DJ, Hill BM, Bailey CP and Ziembicki M (2009) Temporal changes to spatially stratified waterbird communities of the Coorong, South Australia: Implications for the management of heterogenous wetlands. *Animal Conservation* 12(5): 408–417.

Perry N and Gillespie J (2019) Restricting spatial lives? The gendered implications of conservation in Cambodia's protected wetlands. *Environment and Planning E: Nature and Space* 2(1): 73–88.

Pittock J and Finlayson CM (2011) Australia's Murray–Darling Basin: Freshwater ecosystem conservation options in an era of climate change. *Marine and Freshwater Research* 62(3): 232–243.

Ramsar Convention on Wetlands (2018) *Global Wetland Outlook: State of the World's Wetlands and Their Services to People*. Ramsar Convention Secretariat, Gland, Switzerland, 86 pages.

Ramsar Country Profiles (n.d.) *Ramsar Secretariat*. Available https://www.ramsar.org/country-profiles. Accessed 20 October 2019.

Ramsar Site Information Service Site 321 (n.d.) *The Coorong and Lakes Alexandrina and Albert Ramsar Wetlands*. Australia. Available https://rsis.ramsar.org/ris/321. Accessed 26 October 2019.

Ramsar Site Information Service Site 337 (n.d.) *Macquarie Marshes*. Australia. Available http://rsis.ramsar.org/ris/337. Accessed 30 October 2019.

Ramsar Site Information Service Site 1232 (n.d.) *Lake Ngardok Nature Reserve*. Palau. Available https://rsis.ramsar.org/ris/1232. Accessed 24 October 2019.

Ramsar Site Information Service Site 1612 (n.d.) *Upper Navau Conservation Area*. Fiji. Available https://rsis.ramsar.org/ris/1612. Accessed 23 October 2019.

Ramsar Site Information Service Site 2072 (n.d.) *Namdrik Atoll*. Marshall Islands. Available https://rsis.ramsar.org/ris/2072. Accessed 24 October 2019.

Ramsar Site Information Service Site 2331 (n.d.) *Qoliqoli Cokovata Reef*. Fiji. Available https://rsis.ramsar.org/ris/2331. Accessed 23 October 2019.

Richardson BJ (2018) Aesthetics and environmental law: Valuing Tasmania's 'ordinary' nature. *Griffith Law Review* 27(1): 1–30.

Ruddiman WF (2003) The anthropogenic greenhouse era began thousands of years ago. *Climatic Change* 61(3): 261–293.

Sarkkula J, Baran E, Chheng P, Keskinen M, Koponen J and Kummu M (2005) Tonle Sap Lake pulsing system and fisheries productivity. *Internationale Vereinigung für theoretische und angewandte Limnologie: Verhandlungen* 29 (2): 1099–1102.

SPREP (2011) *Regional Wetlands Action Plan for the Pacific Islands*. Secretariat of the Pacific Regional Environment Programme, Apia, 20 pages.

Webster IT (2010) The hydrodynamics and salinity regime of a coastal lagoon—The Coorong, Australia–Seasonal to multi-decadal timescales. *Estuarine, Coastal and Shelf Science* 90(4): 264–274.

Wilson A (2018) *Swamp: Nature and Culture*. Reaktion Books Limited, London.

CHAPTER 6

Conclusion: A Way Forward, Protected Areas and Legal Ecology

Abstract In this final chapter, I reiterate the consistent theme from this book; that a new way of thinking—through the braided interactions of law and geography—can improve the way we execute protected area management. The task has been twofold: (1) To create an applied praxis that directly informs policy makers and (2) To encourage us to approach environmental conundrum (i.e. protected area management challenges) through new eyes with an legal geography lens. In this chapter, I highlight the shortcomings of protected area approaches, for World Heritage and Ramsar listed sites, and reaffirm the value of a legal geography lens through which to work through the complexities of people-place-law dynamics.

Keywords Protected areas · Legal geography · World Heritage · Wetlands · Legal ecology

6.1 Introduction

Law defines, designates and controls human-environment interactions. Halsey (2011, p. 217) suggests

> (L)aw, in short, upholds particular regimes of value (environmental, economic, social, cultural, historical, etc.) which in turn afford various degrees

J. Gillespie, *Protected Areas*,
https://doi.org/10.1007/978-3-030-40502-1_6

of protection (World Heritage area, National Park, conservation park, special protection zone) to particular spaces. In effect, law demarcates the proper and improper use of Nature and is typically taken to occupy a space beyond the damage it is charged with negating.

But what happens when the law gets it wrong, so that neither human values nor non-human values are maintained? This book considers this issue. The essential ingredient for success, it is argued, is that law must be informed by place-people-enviro dynamics. The absence of these in the regulatory recipe dooms it to fail—just like the cake that promises so much as it cooks but falls utterly flat on removal from the oven. If we do not pay attention to these ingredients, carefully accumulating them and weighing them as we mix them and produce the final product, we guarantee it will not work as promised.

6.2 Seeing the Value in Beautiful and Ugly Places

Beauty may be in the eye of the beholder, but does beauty equate to value? In this book we have considered, using selected examples, the state of conservation of very different environmental forms. We first turned to consider how World Heritage sites are protected to benefit all of humanity.

World Heritage properties are some of the world's most famous places. Beyond well-known sites such as Machu Picchu, Athen's Acropolis, Yellowstone, the Great Barrier Reef, Angkor, Easter Island, Venice, the Galapagos Islands, Stonehenge, Cinque Terre, Ha Long Bay, the formal World Heritage list now extends beyond 1000 sites. To make it on to the World Heritage List is a matter of prestige. As the list continues to grow, attention turns to under-represented parts of the world. Gaining a listing comes with obligations to protect and this necessitates an engagement with law (in its various guises). The protection through effective management programs is essential to keeping these sites of outstanding universal value for future generations to enjoy—as mandated through the Convention's text.

A functioning, appropriate regulatory regime for World Heritage properties is not only mandated, it is essential to ensuring their conservation. This necessity draws our attention to the ways in which the regulations are enacted and performed in different contexts. We ask: are the World

Heritage preservation and conservation demands being met? To answer this question, we need to understand what the obligations are. We know from Chapter 4 that the essential regulatory requirements are to: (1) Put in place regulations to protect the property from social, economic and other pressures, (2) To physically demarcate a property boundary and (3) To physically buffer the property through zoning plans. The success of these regulatory efforts is rarely clear-cut. Investigating management techniques, using a legal geography approach, help us shed light on the ways in which these regulatory obligations support (or otherwise) the larger protection objective/s. This is particularly important in the context of the mandated Periodic Reporting targets for World Heritage (and Ramsar) listed properties. Demonstrating whether properties are appropriately conserved using benchmarking methods can be informed using a legal geography approach.

World Heritage properties of outstanding universal value are revered, pedestalled and lauded as remarkable places worthy of conservation across time. Generally, wetlands are not. Infamously the President of the United States, Donald Trump, called for the swamp to be drained, perhaps unaware that Washington, DC. itself was built on the muddy, swamp shores of the Potomac River (Widmer 2017). While draining the swamp in the US context (as in 'cleaning up Washington')—is problematic in a political sense (Friedersdorf 2019), draining swamps in a literal sense is also highly complex.

Wetlands are critically important for planetary health, and provide a raft of important ecosystem services for humans and other living things. Yet, since the emergence and spread of sedentism and agriculture, humans have viewed wetlands with narrow-eyed scepticism bordering on outright loathing. As a consequence, loss of wetland area across the globe, and the attendant habitat fragmentation, loss of biodiversity and ecosystem services, has been prodigious and wetlands are among the most degraded type of environment on Earth. The tension between humanity's ambivalence to these 'ugly' places and our awareness of their importance to us led to the emergence of the Ramsar Convention, arguably humanities largest and most successful international legal instrument for environmental protection. The regulation (read, exclusion or restriction) of human activities in and around wetlands represents a superb example of the disconnect between law's intention and its application, as-it-were, on the ground. In the highly-protected Tonle Sap wetlands of central Cambodia, people in wetland communities describe frustration at the spatial circumstription of their fishing areas and of their fishing techniques, leaving them feeling trapped between rules imposed by 'others' and rapidly changing social and environmental conditions. Under such circumstances, community support for and engagement in protected area management is unlikely. In the Pacific region, individual property ownership within protected wetlands has progressively eroded community-based conservation and weakend the effectiveness of the customary ownership approach to wetland protection. In the absence of strong State control and weakening community control, the failure to articulate development outcomes with conservation goals has eroded any incentive to conserve wetlands, resulting in habitat fragmentation and loss. We also saw in Southeast Asia and the Pacific, unexpected social and even cultural outcomes that emerged from the imposition of in situ protected area regulation. In Australia, the legacy of twentieth century engineering and the grinding realities of twenty-first century hydroclimate variability have exposed profound and dissoluble tensions between wetlands and consumptive users. Even with strong State control, long-established cooperative governance arrangements and a water trading market that sets aside water for environmental purposes, Australia's largest river basins and its internationally protected wetlands are profoundly degraded, resulting in calls for a more adaptive and flexible regulatory approach to protected area management (see, for example, Pittock and Finlayson 2011).

6.3 Reiterating the Importance of Protection Regimes

Whether we are revelling in a majestic World Heritage place or stuck in the quagmires of a swamp there is little doubt about the need, both anthropocentric and ecocentric, to preserve particular lands and waterscapes. The protected area imperative is here to stay, although the specifics of the approach will change form and, hopefully, become more tailored from a management perspective. Under global commitments, namely the Convention on Biological Diversity, and its associated Aichi Targets, we have undertaken to increase the amount of terrestrial and water scapes that are covered by protected area regulatory regimes. Policy reform in protected area management see us moving to visions of protection with inclusive and integrative approaches and with concepts such as equity and justice embedded in policy frames, has become de rigour. This policy shift to incorporate multi-stakeholder perspectives and an inclusive agendum warrants that we pay greater attention to place. Understanding the rigid and contingent people-place-law dynamics becomes an essential component to improving current approaches and to enable us to reach and meet the demands of a new regime. One answer, it is argued, lies in adopting a legal geography approach.

The focus on Target 11 of the Aichi Targets from Chapter 2, means we must increase the area of protected areas while simultaneously ensuring effective and equitable management. This is incredibly challenging and pushes management concerns to the forefront in the fight to maintain ecological and landscape integrity and to stem biodiversity loss. It also brings people and their management through law into sharp focus, and emphasises the importance of analysing the link between the substantive and processual character of law and the everyday lived experience of those impacted by those laws. Law-place-people dynamics lie at the centre of the in situ conservation agenda. This demands the scalar and spatial thinking characteristic of geography must be paired with attention to the nuance of law and law making. It also demands both empirical field-based method and avoiding abstraction of people from the protected area picture. Critically, this understanding is required at, or ideally before, the regulatory drafting stage, not merely as part of a monitoring or periodic reporting program once the laws are enacted. Legal Geography's central attention to place and people gives contextual flexibility—it recognises difference rather than anticipating or even forcing homogeneity. It is thus more able to recognise and accommodate non-Western concepts about land, species, property and conservation and thus facilitate more adaptive, nuanced and ultimately successful law making in a diversity of contexts.

6.4 (Re)New Vision: Towards a Legal Ecology

To understand land and water scapes as interconnected people-place-regulatory systems does mean we need to embrace complexity. We ought not be frugal in our approach through narrow disciplinary lenses but lift our gaze towards understanding the interconnectedness of protected area landscapes.

Two elements intertwine in this book. The first is a pragmatic, policy-oriented praxis aiming at re-orientating decision makers towards a more holistic approach in thinking through protected area programs to enhance their effectiveness and achieve more sustainable solutions for human and non-human world. The second element looks towards a re-conceptualisation of some legal geography scholarship as 'legal ecology'. This is not a claim to re-theorize legal geography (though some might suggest this is necessary) but to re-conceptualise the legal geography project by drawing on and integrating inter-related tools associated with the cognate fields of political ecology and environmental/ecological justice.

I also want to pause to reflect on the production of knowledge. Thinking carefully and critically about situated knowledge production requires that we become circumspect about our research design. This implicates methodology. For insights into the legal geography elements pertaining to methodology see the various works within a dedicated collection on this issue (O'Donnell et al. 2020). In this edited collection a range of legal geography scholars draw out methodological challenges and lessons they have faced in undertaking research in this field. It becomes clear from these collections, and other legal geographer scholarship (Chapter 3), that this perspective encourages an engaged and critical reflection about the meaning of 'equity' and 'justice'. Both law and geography research and scholarship reflect these concerns. Law is very much implicated in giving meaning to these concepts or ideals. Through the application of law we often look to advance justice and equity claims. Putting the regulatory approach under a microscopic lens to understand how law operates to meet these nebulous, often intangible objectives is very much needed both within and outside of academic scholarship. Many, many non-government organisation, business and government agencies are also on-board a paradigm shift that engages with equity and justice as foundational principles for conservation policy and practice (see for example the work of the ICCA Consortium).

Research enables us to better understand our complex world. In uncertain times we look for answers in our research. Shocking biological diversity losses over the past century have led to solutions through conservation programs. To this end in situ protected area policies aimed at preserving habitats and ecosystems to stem biodiversity loss are critical. A number of different types of efforts aimed at improving the protected area effectiveness are taking place. This book is just one part of this larger effort. In it I have sought to shine a light on the place-law dynamics which underpin successful protected area management. To this end, an approach which intertwines these concerns and focusses attention on the place-law imbrication is essential. This gives voice to the call for research to embrace a legal geography lens; to consider a new legal ecology for protected area management in an era of increasing environmental change.

References

Friedersdorf C (2019) You can't drain the swamp and also defend the president. *The Atlantic*, 4 October. Available https://www.theatlantic.com/ideas/archive/2019/10/trump-making-swamp-worse/599344/. Accessed 5 November 2019.

Halsey M (2011) Majesty and monstrosity: Deleuze and the defence of nature. In: Philippopoulos-Mihalopoulos A (ed) *Law and Ecology: New Environmental Foundations*. Routledge, London and New York, pp. 214–236.

ICCA Consortium (undated) *Indigenous Peoples' and Community Conserved Territories and Areas (ICCA)*. Available https://www.iccaconsortium.org/. Accessed 12 November 2019.

O'Donnell T, Robinson DF and Gillespie J (eds) (2020) *Legal Geography: Perspectives and Methods*. Routledge, London and New York.

Pittock J and Finlayson CM (2011) Australia's Murray–Darling Basin: Freshwater ecosystem conservation options in an era of climate change. *Marine and Freshwater Research* 62(3): 232–243.

Widmer T (2017) Draining the swamp. *The New Yorker*, 19 January. Available https://www.newyorker.com/news/news-desk/draining-the-swamp. Accessed 14 November 2019.

Index

J. Gillespie, *Protected Areas*,
https://doi.org/10.1007/978-3-030-40502-1

The manufacturer's authorised representative in the EU is Springer Nature Customer Service Centre GmbH, Europaplatz 3, 69115 Heidelberg, Germany. If you have any concerns regarding our products, please contact ProductSafety@springernature.com

Printed and bound by CPI Group (UK) Ltd, Croydon, CR0 4YY

30/09/2026

02225475-0008